Topographical and Pathotopographical Medical Atlas of the Head and Neck

Scrivener Publishing
100 Cummings Center, Suite 541J
Beverly, MA 01915-6106

Publishers at Scrivener
Martin Scrivener (martin@scrivenerpublishing.com)

Phillip Carmical (pcarmical@scrivenerpublishing.com)

Topographical and Pathotopographical Medical Atlas of the Head and Neck

Z. M. Seagal

Scrivener Publishing

WILEY

Co-published by John Wiley & Sons, Inc. Hoboken, New Jersey, and Scrivener Publishing LLC, Salem, Massachusetts. Published simultaneously in Canada.

For general information on our other products and services or for technical support, please contact our Customer Care Department within the United States at (800) 762-2974, outside the United States at (317) 572-3993 or fax (317) 572-4002.

Wiley also publishes its books in a variety of electronic formats. Some content that appears in print may not be available in electronic formats. For more information about Wiley products, visit our web site at www.wiley.com.

For more information about Scrivener products please visit www.scrivenerpublishing.com.

Cover image is courtesy of the author, Z. M. Seagal
Cover design by Kris Hackerott

Library of Congress Cataloging-in-Publication Data:

ISBN 978-1-119-45973-6

10 9 8 7 6 5 4 3 2 1

Contents

Part 2

About the Author

Z. M. Seagal is an Honoured Scientist of the Russian Federation and the Udmurt Republic, Honorary Academician of the Izhevsk State Medical Academy, Head of the Department of Operative Surgery and Topographic Anatomy, Doctor of Medical Sciences, and Professor.

Topographical and Pathotopographical Medical Atlas of the Head and Neck.

The atlas presents topographical and pathotopographic human anatomy for both the adult and the child. The section "Head" includes cerebral and facial parts according to the areas. Layer topographic anatomy, variant, computer and MRI-topographic anatomy are presented in the section. The surgical anatomy of congenital malformations of the head includes cerebral hernia and hydrocephalus, and of the face - macrostomy, coloboma, clefts of hard and soft palate. In the section "Neck" there are given individual and age differences, fasciae and cell spaces, triangles and vascular-neural bundles, collateral blood supply of the brain in injuries or occlusion of the main neck arteries, topography of the neck organs. All the pictures are colorful and original. The atlas is written in accordance with the educational program of medical universities of the Russian Federation. The original graphs of logical structures are presented according to the sections of topography and congenital malformations. This allows an effective study of the subject.

This atlas is intended for the students of General Medicine, Pediatrics and Dentistry faculties, as well as for interns, residents, postgraduate students and surgeons.The monograph is intended to be used by physicians, junior physicians, medical residents, lecturers in medicine, and medical students.

Introduction

This color atlas of topographic and pathotopographic human anatomy is the fundamental and practically important book designed for doctors of all specializations and students of medical schools. Here you can find almost everything that is connected with the topographic and pathotopographic human anatomy, including original graphs of logical structures of topographic anatomy and development of congenital abnormalities, topography of different areas in layers, pathotopography, computer and magnetic resonance imaging (MRI) of topographic and pathotopographic anatomy. Also you can find here new theoretical and practical sections of topographic anatomy developed by the author himself which are published for the first time. They are practically important for mastering the technique of operative interventions and denying the possibility of iatrogenic complications during operations.

All the pictures are completely original, drawn and colored at a highly professional level. There are also illustrations of individual and age differences of human organs in the chapter of variant topographical anatomy. In the chapter of topographical anatomy you can find holothopy, sceletothopy, synthopy of organs and main anatomical formations, facial vaginae, reticular spaces, branches of nerves and vessels, collateral vessels, zones of sensory and motoric innevation of nerve trunci. Pathotopographic anatomy is displayed by the examples of typical pathology of different areas.

One of the distinguishing features of this book is that topographic and anatomic structures are presented both in corpses and postoperative patients. This data is gathered by the use of MRI and transillumination of organs and tissues.

Unfortunately, nowadays there is no publication of modern atlases of clinical, topographic and pathotopographic anatomy due to several reasons. Firstly, this subject is separated only in Russia and Hungary. In other countries, it is either absent or exists as a part of similar subjects like anatomy or surgery. Some of the themes presented in this book like pathotopographic, variant, normal, topographic anatomy, transillumination, pathotopographic and topographic anatomy of patients and healthy people are original studies of the author of this book.

This color atlas can serve as a textbook for routine necessities of theoretical and practical work of physiologists and pathophysiologists, anatomists and topographic anatomists, pathoanatomists, therapists, surgeons, gynecologists and obstericians, neurologists and dermatovenerologists, radiologists, reanimatologists, anaestesiologists, otorhinolaringologists, traumatologists, orthopedists, pediatricians, dentists and other doctors of narrow specializations like angiosurgeons, urologists, plastic surgeons, neurosurgeons, etc.

Topographic and pathotopographic anatomy presented in this book can be divided into surgical anatomy, therapeutic anatomy, stomatological anatomy, pediatric anatomy, etc. There are also some specific and non-specific indications of pathotopography and variants of physiology, differences between physiology and pathology, and ways of using ultrasound topographic and pathotopographic anatomy. It should be pointed out that studying a living human organism in physiological and pathological conditions is the most effective way for reliable diagnostics. Here you can also find descriptions of less traumatic and non-traumatic operations, as well as effective ways of treatment.

Topographic clinical anatomy changes in the course of a disease and recovery, iatrogenic manipulations, as well as therapeutic and surgical interventions. Some of these situations are described in this book. The author of this book will eagerly accept all constructive criticism and take it into consideration in the next editions of the book.

Part 1: The Head

Topographic Anatomy of the Head

Limits, outer orienteers. The head is separated from the neck with the line which begins on chin elevation – protuberantia mentalis, after that it laterally continues across the lower mandible edge, continues by the lower semicircle of outer aural meatus, goes on to the upper nuchal line, linea nuchae superior, and ends on both sides on outer elevation of occipital bone with protuberantia occipitalis externa. In general, the head is divided into cerebral cranium – cranium cerebrale and facial cranium – cranium faciale.

Cerebral Cranium

Limits: The cerebral cranium is separated from facial cranium with the anatomical formations described below. It is limited by glabella by the middle line. Afterwards this line goes along the brow arch, arcus superciliaris, then - by the rear edge of the processus zygomaticus ossis frontalis and processus frontosphenoidalis ossis zygomatici, then the line goes on by the zygomatic arch, then it continues vertically to the rear edge of the ascending branch of the mandible, and afterwards it continues on the bottom margin where it ends after connecting with a similar line coming from the opposite side – forming protuberantia mentalis.

Area division: The cerebral cranium can be divided into the cranium base, basis cranii, and the cranium fornix (or dome) which is also called calvaria. The cranium base is divided into inner one (basis cranii interna) and external one (basis cranii externa). Calvaria is divided into frontal, parietal, occipital, temporal and mastoid regions (regg. Frontalis, parietalis, occipitalis, temporalis et mastoidea).

Regio frontalis (or the frontal area) is located within the frontal region. Its limiting line begins in the lower part of glabella, spreads to the sides across eyebrow arches, crosses the zygomatic process of the frontal bone, continues upwards across temporal line, and then ascends by the projection line of the coronal suture.

Regio parietalis (parietal region). Limits: anterior limit is the coronal suture, lambdoid suture is the rear limit, temporal line limits it from the sides.

Regio occipitalis (occipital region) is located within the squama of the occipital bone. Limits: lambdoid suture limits it from the top and from the sides and the line drawn horizontally from one apex of mastoid bone to another limits it from below. It's called linea bimastoidea.

Regio temporalis (temporal region). Limits: upper and rear limit is the temporal line, lower limits are zygomatic arch and temporal line above the external aural canal, anterior limit is the zygomatic process of the frontal bone and the outer section of the temporal line. Layer topography of this area has its own complications: skin has less connective tissue septa the lower it is. Aponeurotic helmet becomes thinner and is called here fascia superficialis or superficial fascia. Fascia temporalis propria or proper temporal fascia is formed by the arcus zygomaticus after attaching with its lower margin. You can find its spatium interaponeuroticum, interaponeurotical space, between the plates.

Regio mastoidea or mastoid is an area which is limited with mastoid process. Layer topography of this area has its own complications: skin has no hair and is tightly attached with the mastoid aponeurosis or aponeurosis mastoideus lying below, which is a continuation of galea aponeurotica, but is significantly thinner that its origin. Mastoid process is covered with a thick layer of the periosteum and has attachment points of m. sternocleidomastoideus, mm. splenius capitis et cervicis, venter posterior m. digastrici.

Layers of Regions and their Characteristics:

1. Derma – thick, hair-covered skin, it has a large number of sebaceous and sweat glands. It is firmly attached to the deeper lying subcutaneous fat and

supracranial muscle, musculus epicranius, using vertically going connective tissue septa.

2. Panniculus adiposus – it's subcutaneous fat. Blood and lymphatic vessels of this layer are firmly fixed using connective tissue septa. That's why vessels do not collapse when injured.

3. M. epicranius – it's epicranius muscle which consists of the frontal part, m. frontalis, and the occipital part of the m. occipitatis. Both parts have shared tendon helmet, galea aponeurotica, which is located in the parietal region and becomes thinner as it goes laterally during transition to the temporal region. These three layers which were described above form a single layer of tightly interconnected tissues.

4. Stratum fasciaie – it's fascial layer, a layer of incoherent connective tissue.

5. Periosteum is firmly attached to the skull bones only in the sutural areas.

6. Ossa cranii are the bones of the skull of variable thickness. Occipital region bones have the maximum thickness, and the temporal bones have the minimum thickness. The skull bones are composed of three layers. It's the lamina externa or the external plate, substantia diploica or the spongeous matter which has a lot of vessels – especially vv. diploicae, and lamina vitrea or glass-like plate.

7. Spatium epidurale – it's the space above the dura mater.

8. Dura mater is composed of dense fibrous connective tissue. As it goes deeper, this tunica forms the falx cerebri or the crescent of the brain and with its help it divides cerebral hemispheres from each other; falx cerebelli or the crescent of the cerebellum divides the cerebellar hemispheres, and tentonum cerebelli or the tent of the cerebellum separates the occipital lobes of the cerebral hemispheres from the cerebellum. These processes of dura mater form channels for the drainage of venous blood from the cranial cavity which are called venous sinuses.

9. Spatium subdurale is a space beneath the dura mater.

10. Arachnoidea encephali or the arachnoid membrane contains blood vessels and continues from one gyrus to another without going in.

11. Spatium subarachnoidaie or subarachnoid space stores cerebrospinal fluid in its subarachnoid cisterns or cisternae subarachnoidaie.

12. Pia mater or the soft tunica contains a large number of vessels. The tunica finds its way inside all the sutures between the gyruses.

13. Spatium epicerebrale or epicerebral space. Has the most amount of cerebrospinal liquid between all the listened above spaces.

14. Gyri cerebri or brain gyruses.

Regio Frontalis Has the Next Blood Vessels (Fig.1):

1. a. frontalis is the final branch of the a. ophthatmica and leaves the cranial cavity through the frontal hole, for. frontale. A. frontalis anastomoses with the a. angularis in the corner of the eye;
2. a. suproorbitalis leaves the cranium through the incisura supraorbitalis. Frontal artery lies a little bit more medially than supraorbital. Venous backflow goes through the frontal veins, vv. frontales and after that backflow continues through v. angularis and afterwards it goes on to the system of frontal facial vein. A small amount of blood keeps going to v. ophthatmica superior which pours into the cavernous sinus.

Innervation: motor nerves of this region descend from ramus frontalis n. facialis; sensitive nerves – from ramus ophthalmicus n. trigemini. Lymph drainage of this area is performed through nodi auriculares anteriores, which are located a little a bit more forward than the tragus of the ear.

Regio parietalis is supplied with blood with r. parietalis a. temporalis superficialis. Venous backflow goes through r. parietalis v. temporalis superficialis. Innervation of the area: front is innervated with n. supraorbitalis and n. frontalis, sides – with n. auriculotemporatis, rear – with n. occipitalis major. Lymphatic backflow is supplied with nodi auriculares posteriores.

Regio occipitalis is supplied with blood through a. occipitalis, which branches off a. carotis externa. A. occipitalis gets under m. sternocleidomastoideus and comes close to v. jugularis

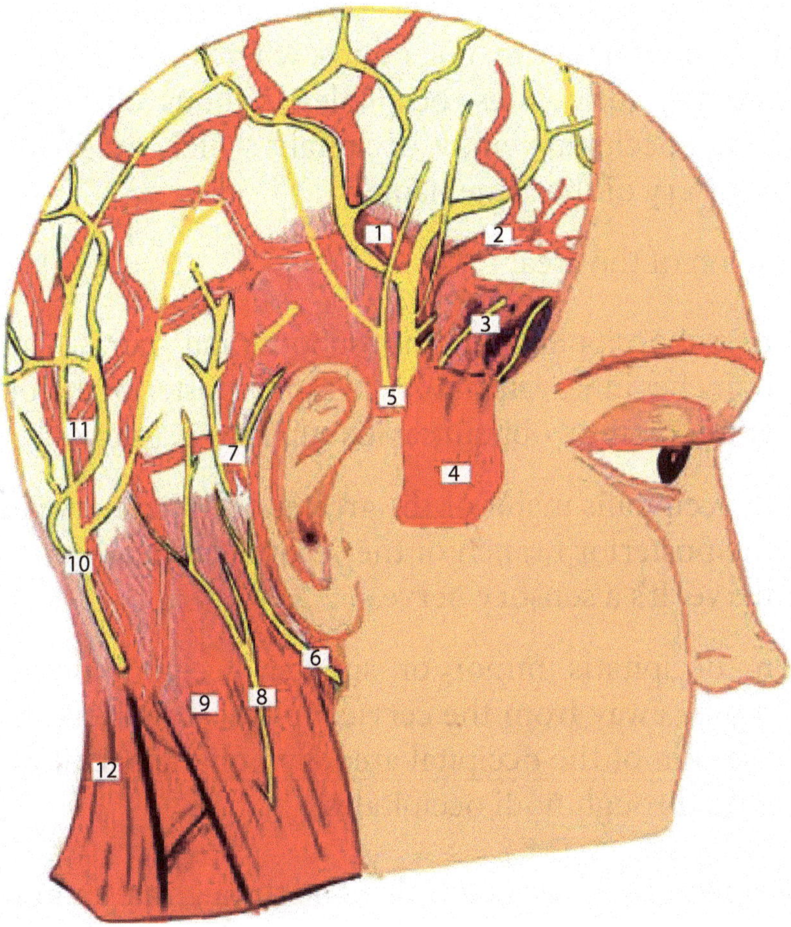

Figure 1 Vessels and nerves of the head.

1 – r. parietalis a. temporalis superficialis, 2 – r. frontalis a. temporalis superficialis, 3 – aa. et nn. temporales profundus, 4 – m. temporalis, 5 – n. auriculotemporalis, 6 – n. auricularis posterior, 7 – a. auricularis posterior, 8 – n. occipitalis minor, 9 – m. sternocleidomastoideus, 10 – n. occipitalis major, 11 – a.occipitalis, 12 – m. trapezius

interna and n. occipitalis minor. Terminal branches a. occipitalis are located under m. epicranius, where they diverge into rr. mastoidei, auriculares, occpitales. Venous outflow goes through the occipital vein, v. occipitalis which runs together with the artery of the same name.

Innervation of the area:

1. n. suboccipitalis or the suboccipital nerve. Its branches innervate mm. recti capitis posterior major et minor, mm. obliqui capitis superior et inferior.
2. n. occipitalis major or the greater occipital nerve is the posterior branch of the second cervical spinal nerve. It's a sensory nerve.
3. n. occipitatis minor or smaller occipital nerve moves away from the cervical plexus to innervate the skin of the occipital area. Lymphatic backflow goes through nodi occipitales.

Regio temporalis is supplied with blood through aa. temporales anterior et posterior (anterior and posterior temporal arteries). Lymphatic vessels traverse into the fossa pterygopalatina and fall into a deeper facial lymph nodes, or nodi faciales profundi. Below it you can find a similar space - the infratemporal fossa or fossa infratemporalis, which communicates with the frontal fossa pterygoidea. There you can run into: a. et v. temporales superficiales, which are located a bit more anterior than the tragus of the ear. The vein itself is placed behind the artery. Also there you can find n. auriculotemporalis which separates from the third branch of the trigeminal nerve, n. zygomaticotemporalis, which comes out of the cheek holes of the temporal bone.

This nerve is a branch of the zygomatic nerve. Both nerves innervate the skin of the temporal region.

Regio mastoidea: Blood supply and venous drainage is carried out by a. et v. auricularis posterior. Innervation is performed via n. auricularis magnus and n. occipitalis minor.

Cellular spaces. The temporo-parietal-occipital region (Fig. 2) has the subaponeurotic space between the tendon and the periosteum helmet, which is locked with the cojoinment of the helmet with the periosteum. Subperiosteal space is confined within a single bone. In the temporal region (Fig. 3) cellular tissue has four layers. There is fat between the skin and superficial fascia (panniculus adiposus).

Interfascial temporal space, spatium temporale inteifasciale, is located between the superficial and deeper temporal fascia plates; it also contains adipose tissue. Cellular tissue also can be found between superficial and deep temporal fasciae and the temporalis muscle. Process of cheek fat body (corpus adiposum buccae) enters this cellular tissue.

Musculoskeletal temporal space is located between the inner surface of the temporal muscle and periosteum. Below this space communicates with the temporopterygoid space.

The structure of the bones of the skull and mastoid process in adults and children. The skull bones (ossa cranii) have different thickness at different areas. The smallest thickness is located in the temporal region within the squamous part of the temporal bone (pars squamosa ossi temporalis), while the thickest area is found in the occipital region. The bones of the skull have a major difference when compared with the rest of the bones of the skeleton: they do not regenerate. Bones

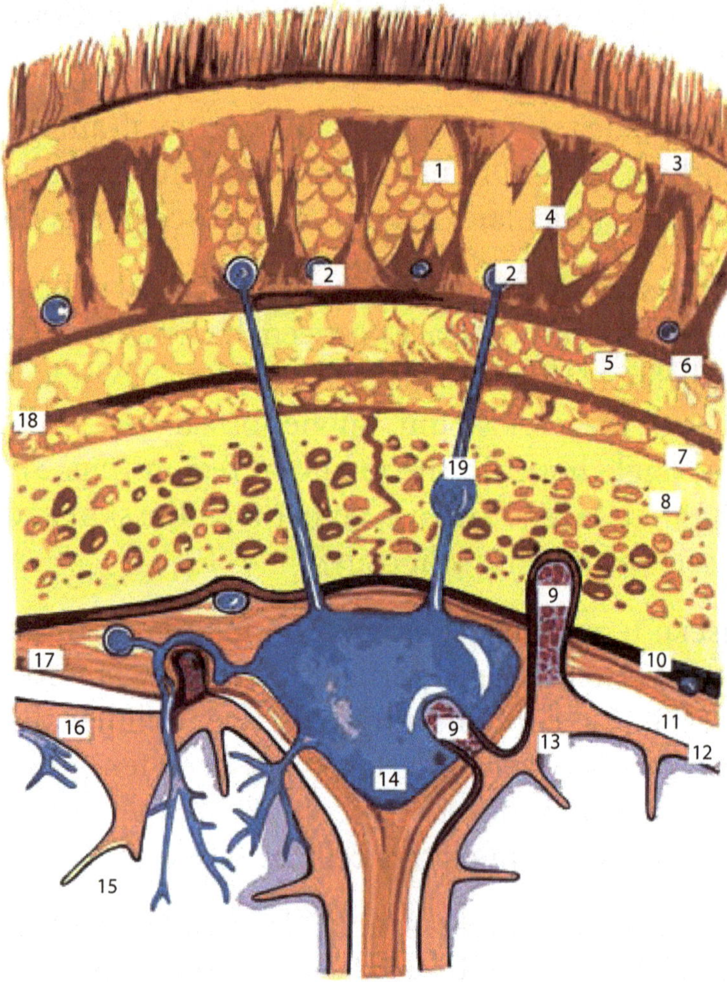

Figure 2 Layers of the frontal-parietal-occipital area.

1 – panniculus adiposus, 2 – vessels of subcutaneous layer, 3 – derma, 4 – tendon intersections 5 – subaponeurotic cellular tissue, 6 – aponeurosis, 7 – subperopostal cellular tissue, 8 – os parietalius, 9 – Pacchioni's granulationes, 10 – cavitas epiduralis, 11 – spatium subduralis, 12 – tunica arachnoidea, 13 – tunica vasculosa, 14 – sinus venosum, 15 – cerebrum, 16 – spatium subarachnoidalis, 17 – dura mater encephali, 18 – periosteum, 19 – vv. Diploicae

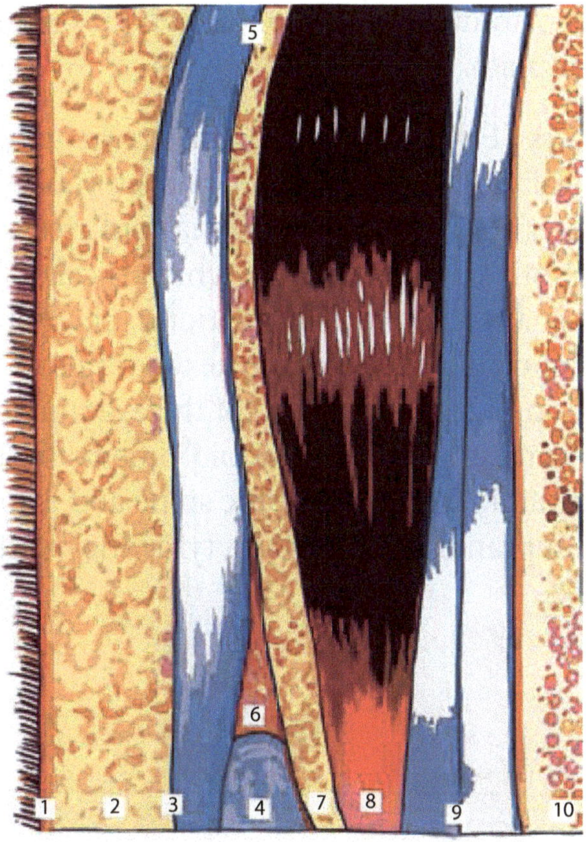

Figure 3 Regio temporalis.

1 – derma, 2 – panniculus, 3 – fascia superficialis, 4 – arcus zygomaticus, 5 – poneurosis temporalis, 6 – interaponeurotic cellular tissue, 7 – subaponeuritic cellular tissue, 8 – m. temporalis, 9 – periosteum, 10 – os temporale.

of the skull have three layers: 1) The outer plate (lamina externa) consists of a compact matter about 1 mm thick. 2) spongeous matter (diploe) contains a large amount of diploic veins. 3) inner (vitreous) plate or lamina interna (s. vitrea) also is made of a compact matter about 0.5 mm thick. On the inner side of the skull you can find sulci arteriosi, which provide the possibility of close contact of dura mater's vessels with the inner plate.

Mastoid process (Fig. 4) (processus mastoideus) has a rough surface, especially on the rear side, where sternocleidomastoideus muscle gets attached. Inside the mastoid area you can run across the smooth triangular shape, which is called the Shipo's triangle. It's usually used for trepanations. The boundaries of the triangle Shipo are defined by the next boundary: top limit is the continuation of the upper edge of the zygomatic arch, rear limit is the frontal edge of the mastoid tuberosity (tuberositas mastoidea), front limit is suprameatic spine and drum-mastoid fissura (spina suprameatum et fissura tympano-mastoidea).

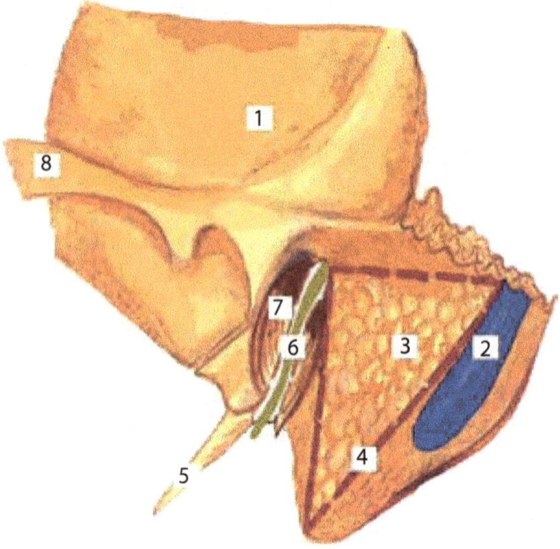

Figure 4 Mastoid process with Shipo's triangle.

1 – squama ossis temporale, 2 – the projection of the sigmoid sinus, 3 – projection of the mastoid cells, 4 – Shipo's Triangle 5 – processus styloideus, 6. projection of the facial nerve canal, 7 – meatus acusticus externus, 8 – arcus zygomaticum.

Cranial of the newborn. Neonatal cranial bones are thin, easy to bend, and are connected together with fibrous membranes. They can shift comparatively to each other. As a result, the head can change its shape when pressure is applied which is critical when the child moves though birth passway.

Fontanelles are characteristic features of children's skulls. There is:

1. anterior fontanelle (fonticulus anterior) which is located at the junction of the parietal and frontal bones. It has a rhombic shape and usually closes when child is 1–1.5 years old.

2. posterior fontanelle (fonticulus posterior) which can be found at the junction of the parietal and occipital bones. It usually closes by the end of first 2 months of life.

3. sphenoid fontanelle (fonticulus sphenoidalis) which is located at the junction of the parietal, frontal, sphenoid and temporal bones. It closes at the end of the prenatal period or shortly after birth.

4. mastoid fontanelle (fonticulus mastoideus) is located at the junction of the temporal, parietal and occipital bones and closes at the end of the prenatal period or in the first few months after birth.

External and internal base of the skull. At the base of the skull inner surface you can find basis cranii interna (Fig. 5) and an outer surface – basis cranii externa.

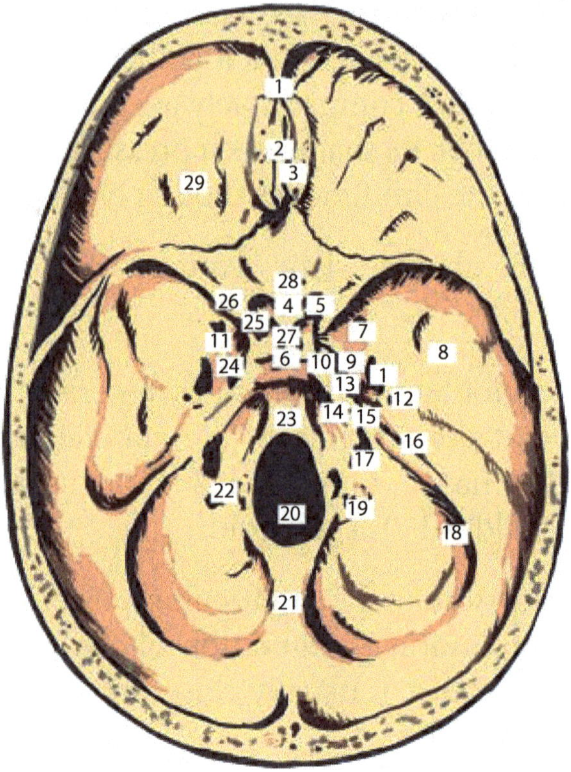

Figure 5 Inner cranium base.

1 – for. caecum, 2 – crista galli, 3 – lamina cribrosa, 4 – sulcus chiasmae, 5 – canalis opticus, 6 – dorsum turcicae sellae, 7 – for. rotundum, 8 – ala major ossis sphenoidalis, 9 – for. accessorius, letting vein through from a net of foramen ovale 10 – processus sphenoidalis posterior, 11 – for. ovale, 12 – for. spinosum, 13 – for. trigeminum, 14 – sulcus sinus petrosi inferiors, 15 – margo tubae auditivae internae, 16 – sulcus sinus petrosi superiors, 17 – for. jugulare, 18 – sulcus sinus sigmoidei, 19 – margo canalis condylaris, 20 – for. occipitales maior, 21 – crista occipitalis interna, 22 – margo canalis hypoglossalis, 23 – clivus, 24 – for. lacerum et bone bridge separating foramen into two parts 25 – processus spenoidalis anterior, 26 – ala minor ossis sphenoidalis, 27 – fossa hypophisialis, 28 – tuberculum sellae, 29 – pars orbitalis ossis frontalis.

Basis Cranii Interna

Inside the fossa cranii anterior you can take a look on:

1. lamina cribrosa ossis ethmoidalis – ethmoid plate of the ethmoid bone. Olfactory threads (filae olfactoriae which belong to the first pair of cranial nerves) go through its holes;

2. foramen coecum is located forward from the crista galli. It links with the nasal cavity with venous anastomoses between the nasal cavity and the superior sagittal sinus. Impressiones digitae are the result of brain gyruses and cranium developing in close contact. They are separated from each other with brain elevations, juga cerebralia;

3. canalis opticus or optical channel. Inside it you can find the II pair of cranial nerves and the ophthalmic artery, a. ophthalmica. During aneurisms of a. ophthalmica aneurismic sac presses onto optical nerve, which can result into amblyopia (degraged vision) or amauvrosis (blindness).

Inside the middle cranial fossa, which is anteriorly limited with the smaller wings of the sphenoid bone, posteriorly – with the pyramid of the temporal bone and partially also with the Turkish saddle, you can locate:

1. fissura orbitalis superior – upper eye fissure. V. ophthalmica superior, III, IV, VI pairs of brain nerves and following sensory nerves: nn. frontalis,

lacrimalis, nasociliaris, which are branches of r. ophthalmicus n. trigemini – they all go through it.

2. foramen rotundum - or a round hole. It opens into fossa pterygopalatina and contains r. maxillaris n. trigemini;
3. foramen ovale - oval hole. It opens into the lower temporal fossa and contains a. maxillaris, r. mandibularis n. trigemini;
4. foramen spinosum - spinous hole. There are a.meningea media and r. meningeus n. mandibularis;
5. foramen lacerum - ragged hole. There are: nn. petrosi majoris et minoris, tuba auditiva, m. tensor tympani and n. tensor tympani;
6. foramen caroticum internum - inner carotic hole. A. carotis interna with plexus sympaticus are passing through it;
7. hiatus canalis facialis - the opening of the facial canal. N. petrosus major, ramus petrosus a. meningeae mediae, v. auditiva proceed through it.
8. apertura superior canaliculi tympanici - the top opening of the drum tubule. N. petrosus minor and a. tympanica pass it.

Posterior fossa is limited by the pyramid of the temporal bone, slope and cruciform elevation, eminentia cruciata. Following formations are located here:

1. foramen magnum - a large hole where. Medulla oblongata, n. accessorius, a. vertebralis, n. spinalis pass through it;

2. foramen jugulare - the jugular hole. IX, X, XI cranial nerves go in the front, v. jugularis interna and a. meningea posterior proceed in the rear;

3. canalis hypoglossi - hypoglossal channel for the nerve with the same name;

4. porus acusticus internus - the internal auditory canal. Inside it you can find n. facialis, n. acusticus, n. intermedius, a. auditiva interna, vv. auditivae internae;

5. apertura externa aqueductus vestibuli - external opening of the aqueduct. Inner lymphatic passage goes across this aperture.

6. apertura externa canaliculi cochleae - external opening of the cochlear canal. V. canaliculi cochleae goes across this aperture;

7. foramen mastoideum - mastoid hole. Emissarium mastoideum, which links sinus sigmoideus with v. occipitalis, can be found inside.

Basis cranii externa. The following formations are defined on the external cranium base:

1) foramen magnum, 2) foramen condyloideum: it serves as a connection between sinus sigmoideus and plexus venosus of the neck area; 3) canaiis hypoglossi with the same-called

nerve 4) foramen mastoideum, 5) foramen stylomastoideum, stylomastoid hole: n. facialis, a. et v. stylomastoidea are located inside; 6) foramen caroticum externum; 7) foramen lacerum; 8) porus acusticus externus; 9) apertura inferior canaliculi tympani, or the lower opening of the drum tubule, where the a. tympanica inferior and n. tympanicus go; 10) foramen spinosum; 11) foramen ovale; 12) canalis pterygoideus, pterygoid canal - nn. petrosus major et profundus.

The Brain

The cerebral hemispheres: the large brain is divided into two hemispheres: the right (hemispherium dextrum) and the left (hemispherium sinistrum). Cerebral hemispheres have superolateral surface (facies superolateralis), the bottom surface (facies inferior) and the medial surface of the brain (facies medialis). Each hemisphere is divided into pallium or cortex, the olfactory brain (rhinencephalon) and basal (subcortical) nucleui lying deep in the medulla hemisphere. Hemispheres are separated by the longitudinal fissure of the brain (fissura longitudinalis cerebri).

Lobes: Each pallium is divided into four lobes: the frontal lobe (lobus frontalis), upper lobe – parietal lobe (lobus parietalis), rear lobe – occipital lobe (lobus occipitalis), lateral lobe – temporal lobe (lobus temporalis).

Frontal lobe. Front has the frontal pole (polus frontalis), rear - with the central sulcus (sulcus centralis), lower part - the lateral sulcus (sulcus lateralis).

Parietal lobe. Front has the central sulcus (sulcus centralis), lower part - the lateral sulcus (sulcus lateralis), rear – an imaginary line drawn between sulcus parietooccipitalis and incisura preoccipitalis.

Temporal lobe. The front has temporal pole (polus temporalis), upper part - the lateral sulcus (sulcus lateralis), rear - an imaginary line drawn between sulcus parietooccipitalis and incisura preoccipitalis.

Ooccipital lobe. Front has an imaginary line drawn between sulcus parietooccipitalis and incisura preoccipital, bottom and rear - occipital pole (polus occipitalis).

Sulci: Central (Rolando's) sulcus - sulcus centralis, which separates the frontal lobe and the parietal. The lateral sulcus - sulcus cerebri lateralis - separates the temporal lobe and the parietal. The frontal lobe has two longitudinal sulci – upper sulcus, sulcus frontalis superior, and lower sulcus, sulcus frontalis inferior. Sulcus interparietalis separates the inferior parietal lobule from the superior one. The parietal region also has a superior temporal sulcus. There is a sulcus in the front of the occipital lobe – it's sulcus parietooccipitalis. Temporal lobe has 3 sulci: the upper one, middle one and lower one -sulci temporales superior, medius et inferior.

Brainstem. It's divided into an intermediate brain, midbrain, pons and medulla oblongata (Fig. 6).

Intermediate brain: it has visual hillocks - thalami optici, pars mamillaris hipothalami, epithalamus, metathalamus. Thalami are divided from surrounding brain tissue with stria terminalis. III brain ventricle lies between those thalami. The

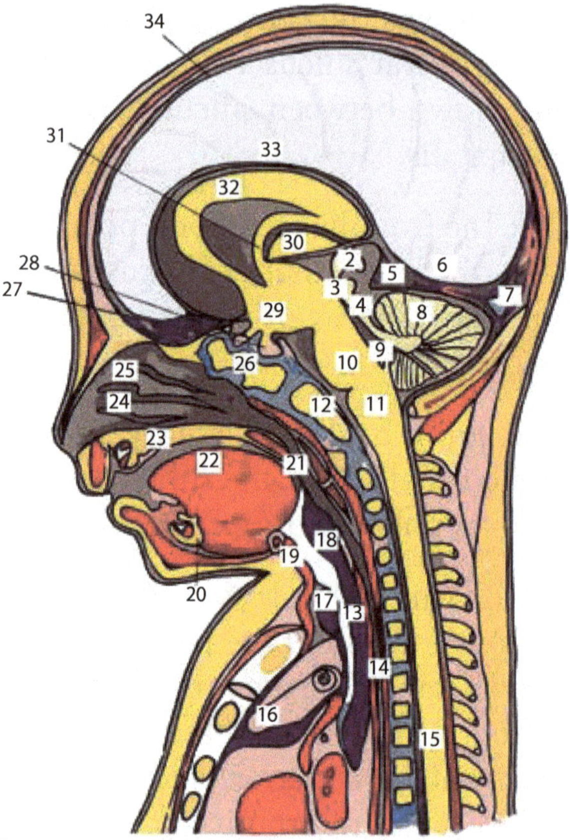

Figure 6 Medial section of head and neck

1 – falx cerebri; 2 – corpus pineale; 3 – aqueductus mesencepali; 4 – tectum mesencepali; 5 – tentorium cerebella; 6 – sinus rectus; 7 – confluens sinuum; 8 – cerebellum; 9 – ventriculus IV; 10 – pons; 11 – medulla oblongata; 12 – pars basilaris ossis occipitalis; 13 – trachea; 14 – esophagus; 15 – medulla spinalis; 16 – thymus; 17 – gl. thyroidea; 18 – cavitas larynges; 19 – os hyoideum; 20 – mandibula; 21 – cavitas pharynges; 22 – lingua; 23 – palatum durum; 24 – concha nasalis inferior; 25 – meatus nasi medium; 26 – corpus ossis sphenoidalis; 27 – hypophysis; 28 – chiasma opticum; 29 – hypothalamus; 30 – thalamus; 31 – columna fornicis; 32 – corpus collosum; 33 – sinus sagittalis inferior; 34 – sinus sagittalis superior.

midbrain (mesencephalon) has the next formations: corpora quadrigemina, pedunculi cerebri, isthmus of rhomboid brain - isthmus rombencephali. Pons (Varolius' Bridge) is a white elevation placed transversally. It has sulcus basillaris on its lower surface. In the front pons is limited with pendiculi cerebri, in the rear with the medulla oblongata. Medulla oblongata: it has an anterior medium fissure in the front and the posterior transversal fussier in the rear.

The pyramid lies on the anterior surface. On the outside the pyramid is limited with the front side sulcus, olive itself lies in the direction to the surface. On the rear surface of the medulla oblongata, by the sides of the medium fissure you can find two sulci: posterior lateral sulcus and posterior intermediate sulcus.

Cranial Nerves:

I. **pair**, n. olfactorius - the olfactory nerve starts from the olfactory field, which is located at the rear of the middle and upper part of nasal choanae. Beginning from there, about 18-20 olfactory filaments penetrate through the ethmoid plate into the anterior cranial fossa, where the nerve enters into the olfactory bulb and then proceeds by the olfactory tract. Higher olfactory centers are gyrus parahippocampalis and uncus. Damage of the first pair of cranial nerves at every level results into disorder of sense of smell, hyposmia, or its loss - anosmia, its strengthening - hyperosmia or distortion of perception, parosmia.

II. **pair**, n. opticus - the optic nerve. It originates from the rods and cones of the retina, then as the optic nerve it proceeds through the foramen opticum. Soon it falls into the cranial cavity. Some of its fibers cross near the sella Turcica, so after that the optic tract goes to the thalamus and the lateral geniculate body. Damage of the optic nerve results into vision degradation (ambiopy) or blindness (amaurosis). With the defeat of the optic tract occurs off a unilateral visual field in both eyes hemianopsia homonima. Damage of the optic chiasm results into disabling inner or outer fields of vision.

III. **pair**, n. oculomotorius - oculomotor nerve. It innervates the upper, inner and lower rectus muscles and the lower oblique musclus levator of the eye and the musculus levator of the upper eyelid. Parasympathetic fibers of the sphincter of the pupil (m. sphincter pupillae) are also part of this pair. Damage to this nerve results into exotropia - strabismus divergens, ptosis or dilation of pupils - mydriasis.

IV. **pair**, n. trochlearis - a block nerve. It innervates the upper oblique muscle eyes. It comes from quadrigeminum and passes through the upper orbital fissure into the orbit. Its damage results into feeling of diverged eyeballs and a sense of diplopia.

V. **pair** (Fig. 7), n. trigeminus - trigeminal nerve. Its motor nucleus is located on the border between

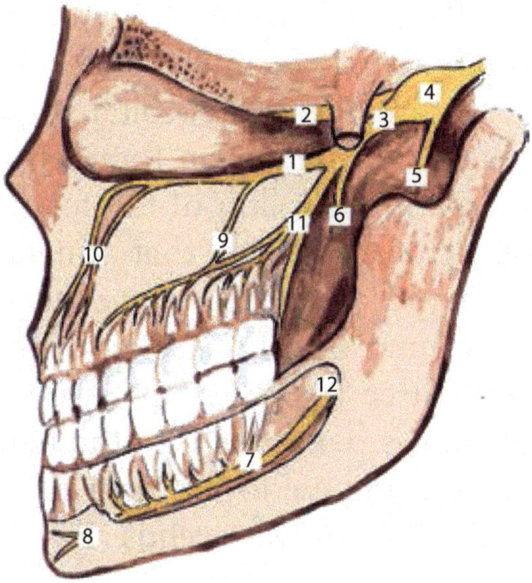

Figure 7 Branches of trigeminal nerve

1 – n. mentalis; 2 – n. orbitalis; 3 –n. maxillaris; 4 – ganglion semilunare (Gasseri); 5 – n. mandibularis; 6 – n. pterygopalatinum; 7 – n. alveolaris inferior; 8 – n. mentalis; 9 – n. alveolaris superior medius; 10 – nn. alveolares superiores anteriores; 11 – n. alveolaris superior posterior; 12 – canalis mandibularis.

the upper part of the rhomboid fossa and the pons. Fibers go out of these cores to the top of the temporal bone pyramid where the trigeminal node can be found in the splitting of the dura mater.

It has three branches:

1. Nervus ophtalmicus - ophthalmic branch. It goes to the upper-orbital hole and before entrance it divides into n. frontalis - frontal nerve, n.

lacrimatis - lacrimal nerve, n. nasolacrimalis - nasolacrimal nerve. Terminal branches of the frontal nerve are the supraorbital branch, that exits through the sulcus or foramen supraorbitaie, innervating the skin of the forehead, and the supratrochlear branch that innervates the skin of the upper eyelid. Lacrimal nerve innervates the lacrimal gland. Nasolacrimal nerve is divided into three branches: the front and rear ethmoidal nerves penetrate into the nasal cavity through nasal foramen; lower block nerve innervates the skin of the lower eyelids (Fig. 8).

2. Nervus maxillaris - maxillary nerve which passes through the round hole of the skull base and divides into the following branches:

- n. zygomaticus - zygomatic nerve, innervates the skin chin area;
- n. intermedius - intermediate nerve, participates in the formation of ganglion pterygopalatinum;
- n. alveolares superiores posteriores are rear upper alveolar branches which pass through foramina alveolaria posteriora on tuber maxillae in the alveolar process to provide the innervation of the back upper teeth;
- n. infraorbitalis - infraorbital nerve. It passes through the infraorbital fissure, then goes through the canalis infraorbitalis and forms a so-called "smaller goose foot", which is located

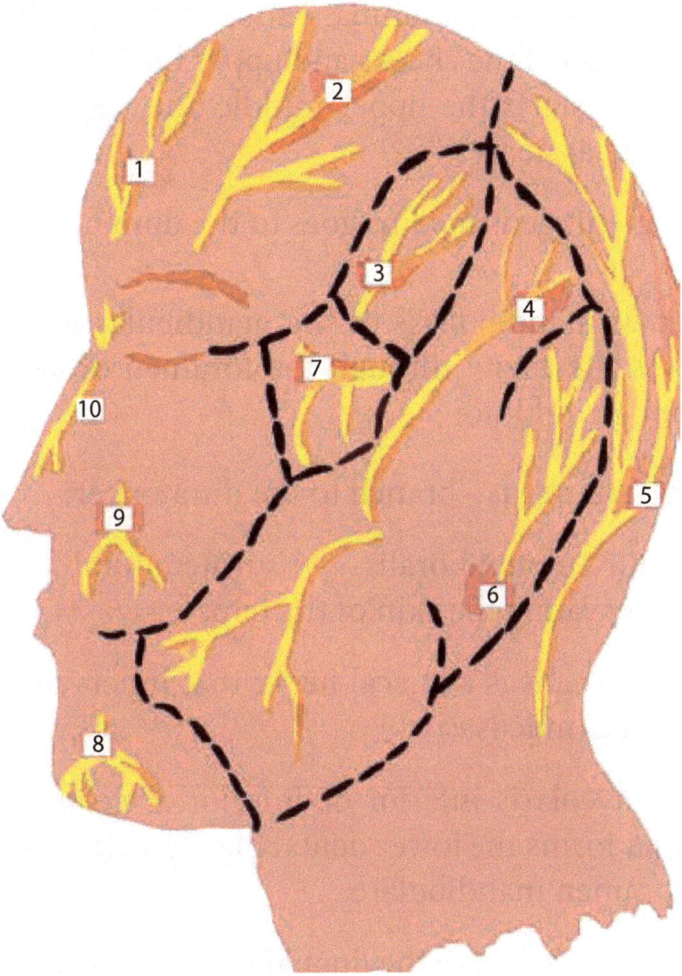

Figure 8 Scheme of head skin innervation.

1 – n. supratrochlearis; 2 – n. supraorbitalis; 3 – n. zygomaticotemporalis; 4 – n. auriculotemporalis; 5 – n. occipitalis major; 6 – n. occipitalis minor; 7 – n. zygomaticofacialis; 8 – n. mentalis; 9 – n. infraorbi-talis; 10 – n. ethmoidalis anterior.

within the fossa canina. “Goose foot” innervates the skin of the cheeks and upper lip. Infraorbital nerve starts the upper middle and upper front alveolar nerves;

- r. meningeus medius goes to the dura mater.

3. Nervus mandibularis or the mandibular nerve. It leaves the cavity through the foramen ovale, and branches off into:

- r. meningeus - branch to the dura mater;
- n.auriculotemporalis - aurotemporal nerve innervating the skin of the temple;
- n. buccalis is a buccal nerve that innervates the buccal mucosa;
- n. alveolaris inferior or inferior alveolar nerve that forms the lower dental plexus after entering foramen mandibulare;
- nn. linguales are the lingual nerves that carry on taste sensitivity from the frontal two-thirds of the tongue;
- n. mylohyoideus is the maxillo-hypoglossal nerve innervating mylohyoid muscle and anterior corpus of digastric muscle;
- nervus massetericus innervates chewing muscles;

- nervi pterygoidei lateralis et medialis attach to the pterygoid muscles;
- rami temporales proceed to the temporal muscle;
- n. musculi tensoris veli palatini innervate the muscle of the same name;
- n. musculi tensoris tympani innervate the muscle of the same name.

Symptoms of trigeminal nerve are hypersthesia, anesthesia or hyposthesia localized in zones of innervation of the trigeminal nerve, paralysis of the masticatory muscles; pain in the exit points of the nerve branches: foramen supraorbitale, foramen infraorbitale, fossa canina and foramen mentale; lockjaw when there is a damage of mandibular branches as a result of hematomic compression hematoma originating from face wounds; trophic keratitis.

VI. **pair** is n. abducens. The core of this motor nerve is located at the bottom of the rhomboid fossa. It passes through fissura orbitalis superior and innervates the lateral rectus muscle of the eye. Symptoms of damage inflicted to VI pair: in case of prevailing of tone of the internal rectus muscle, there is a chance of appearing of an internal strabismus – or strabismus convergens. In case of inflammation of the pars petrosa of the temporal bone (petrositis), which can happen as a consequence of the inflammation of the middle ear, there is a chance of symptoms of the abducens and the oculomotor nerves. This symptom complex is called Gradenigo syndrome - the immobility of the eyeball.

VII. **pair** (Fig. 9,10), n. facialis – the facial nerve. The core of the motor nerve is also located in the rhomboid fossa. It goes through the internal auditory canal, and after that it proceeds into the facial canal of the temporal bone. Within the facial canal this nerve starts following branches:

- n. petrosus major. On the front surface of the pyramid of the temporal bone it lies in the sulcus n. petrosi majoris. Then it passes through foramen lacerum and connects with n. petrosus profundus and participates in the formation of ganglion pterygopalatinum;
- n. stapedius - goes to the tympanic cavity and innervates m. stapedius;
- chorda tympani. Together with the lingual nerve it innervates the frontal two-thirds of the tongue.

After exiting the foramen stylomastoideum facial nerve pierces the parotid-masticatory gland, within which it forms a "big goose foot": rami temporaiis go to the frontal muscle, rami zygomatici – to the circular muscles of the eye, rami buccalis - to the facial muscles persons, ramus marginalis mandibulae – goes on to the edge of the lower jaw and innervates the m. quadratus labii inferior and m. mentalis, ramus colli connects with m. piatysma, ramus auricucularis posterior – with some rudimentary ear muscles, ramus digastricus innervates the rear ventriculus of digastric muscle. Symptoms

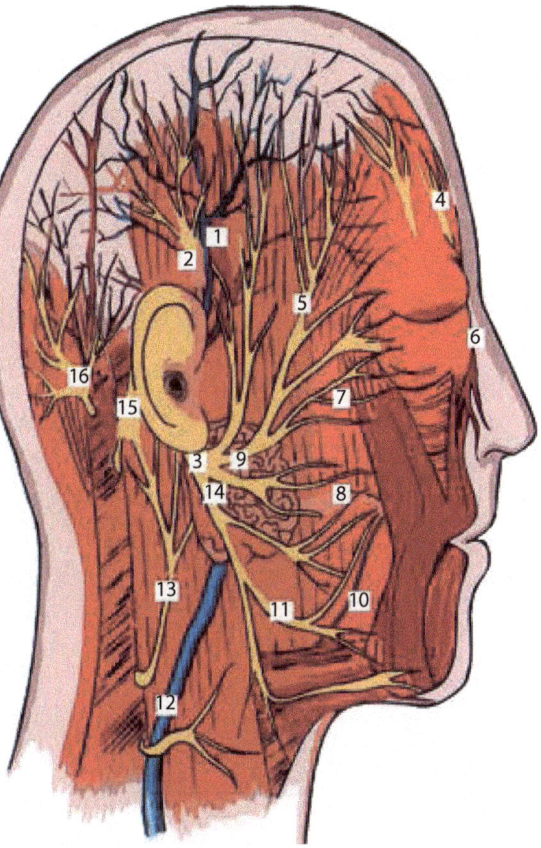

Figure 9 Superficial head formations.

1 - a. temporalis superficialis; 2 - n. auriculotemporalis; 3 - n. frontalis; 4 - n. supratrochlearis; 5 - r. temporalis n. facialis; 6 - a. angularis; 7 - r. zygomaticus n. facialis; 8 - ductus parotidea; 9 - r. buccalis n. fa-cialis; 10 - aa. faciales et vena; 11 - r. marginalis n. facialis; 12 - v. jugularis externa; 13 n. auricularis major; 14 - gl. parotidea; 15 - n. occipitalis minor; 16 - n. occipitalis major et vasae

associated with the facial nerve: Prosopoplegia homolateralis – loss of function of facial muscles on one side with shift to the healthy side. Logophthalmus paralyticus – "never sleeping eyes". This happens because of the dominance of tone m. levator palpebrae superior above m. orbicularis oculi. Loss

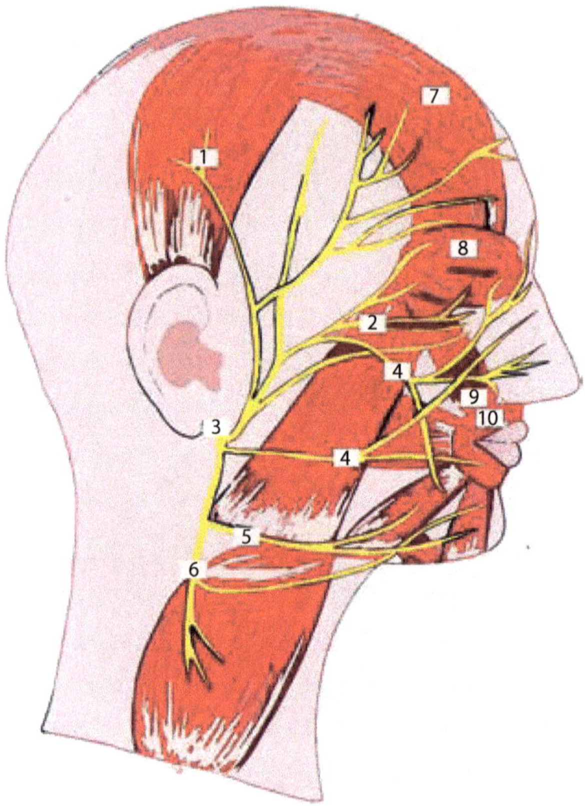

Figure 10 Branches of the facial nerve

1 – r. temporalis; 2 – r. zygomaticus; 3 – truncus n. facialis; 4 – r. buccalis; 5 – r. marginalis mandibulae; 6 – r. cervicalis; 7 – m. frontalis; 8 – m. orbicularis oculi; 9 – m. quadratus labii superior; 10 – m. orbicularis oris.

of taste sensations of the front two-thirds of the tongue. Distorted perception of sounds. Problems with tear glands.

VIII. **pair**, n. vestibulo-cochlearis - vestibular-auditory nerve. It consists of n. cochlearis - cochlear nerve conducting hearing impulses, and n. vestibularis - vestibular nerve associated with maintaining balance. The core of the cochlear nerve is located at the bottom of the rhomboid fossa, and the vestibular nerve nuclei are located at the bottom of the IV ventricle. Symptoms of vestibular-auditory nerve: in case of fracture of the skull base with damage dealt to the temporal bone pyramid all three nerves passing in the inner ear canal are also damaged: n. facialis, n. intermedius, n. acustici - that's why in this case the patient usually loses hearing, starts having problems with salivation, and meets all the difficulties associated with the facial nerve.

IX. **pair**, n. glossopharyngeus - glossopharyngeal nerve. Its core is laid at the bottom of the rhomboid fossa. Nerve exits the cavity through the jugular foramen, right between v.jugularis interna and a. sagotis interna and proceeds to the root of the tongue, where it starts its terminal branches: r. lingualis and rami pharyngei. Immediately upon exit from the jugular foramen it also begins the n. tympanicus.

X. **pair**, n. vagus - the vagus nerve. Its core is also located at the bottom of the rhomboid fossa. After leaving through foramen jugulare, it starts n. laryngeus superior, which innervates the mucous membrane of the larynx and m. cricothyroideus. Below it n. gesugrens takes its departure – its a recurrent nerve that innervates the trachea, esophagus and thyroid gland. Symptoms of the vagus nerve: if it is damaged above the recurrent nerve - disturbance of cardiac activity, respiration, loss of functions of the larynx, as well as its sensitive paralysis.

XI. **pair**, n. accessorius - accessory nerve, it originates from the cervical part of the spinal cord and enters the cranial cavity through the foramen magnum and the jugular foramen through again sent to the neck, which supplies the trapezius and the sternocleidomastoid muscle. One symptom of the accessory nerve is torticollis. If it is damaged at the skull base in the area of jugular foramen it causes syndromes similar to the damage of IX, X, XII pairs of brain nerves.

XII. **pair**, n. hypoglossus – hypoglossal nerve. Its core is located at the bottom of the rhomboid fossa. This nerve goes through the canalis hypoglossus. It innervates the muscles of the tongue. It starts branches of motor innervation of sternohyoid, grudinoschitovidnoy, scapular-hyoid and thyroid muscles. Symptoms of the hypoglossal nerve: tilt of the tongue to the damaged side, atrophy of muscles innervated by n. hypoglossus, oblique arrangement of the larynx.

Tunicas of the brain: 1) dura mater, 2) the arachnoid membrane - tunica arachnoidea, 3) pia mater (tunica vasculosa).

Epidural and intrathecal space: 1) spatium epidurale - space above the dura, 2) spatium subdurale - space underneath the dura, 3) spatium subarachnoidale - subarachnoid space, although sometimes this space forms an extension - subarachnoid cisterns with a large quantity of cerebrospinal liquid. 4) spatium epicerebrale - epicerebral space. All these spaces store cerebrospinal fluid. The biggest amount of this liquid lies in the subarachnoid space.

Specifics of arterial blood supplying and backflow of venous blood from the brain. The blood supply of the brain is carried out using branches of four arteries: the two internal carotid arteries and two vertebral arteries (Fig. 11, 12).

The internal carotid artery (a. Carotis interna) detaches from the common carotid artery at the level of the upper edge of the thyroid cartilage.

Along the way the internal carotid artery starts the following branches: a. ophthalmica, a. lacrimalis, a. centralis retinae, aa. ciliares, rr. musculares, a. ethmoidalis posterior, a. ethmoidalis anterior, a. supraorbitalis, a. supratrochlearis, a. dorsalis nasi, a. temporalis anterior, a. temporalis intermedius, a. temporalis posterior, rr. temporales posteriores, r. calcarinus, r. parietooccipitalis. There are also some smaller branches of coming to the middle front gyrus, a precentral, postcentral gyrus and superior parietal lobule.

Vertebral artery. After ascending up the slope, both vertebral arteries begin lower cerebellar arteries (aa. Cerebelli posterior inferior) and merge into one unpaired basilar artery (a. Basilaris). This artery has the following branches: a. cerebelli inferior anterior, a. labyrinthica, aa. pontis, a. cerebelli superior, aa. cerebri posterior, rr. temporales anteriores, rr. temporales posteriores, r. calcarinus, r.parietooccipitalis. Communicating artery (a. Communicans posterior) goes from the posterior cerebral artery to the internal carotid artery. Arterial circle of the cerebrum, circulus arteriosus

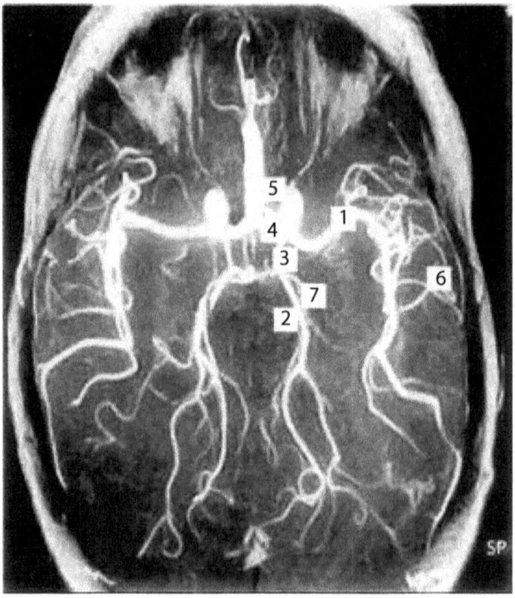

Figure 11 Cerebral vessels. MRI.

1 – a. cerebri media; 2 – a. communicans posterior; 3 – facies medialis; 4 – a. cerebelli anterior; 5 – a. communicans anterior; 6 – facies lateralis; 7 – a. cerebri posterior

cerebri (Willisii) is made of: up front - of the unpaired anterior communicating artery (a. Communicans anterior), from the anterolateral side - of the anterior cerebral artery (a. Cerebri anterior), from the outside - of the posterior communicating artery (a. communicans posterior), from behind - of the posterior cerebral artery (a. cerebri posterior).

Cerebral veins are arranged in three layers. The most superficial veins are located in the soft tissues of the head. The flat bones of the skull are supplied with diploic veins (vv. Diploicae). Inside the cranial cavity blood flowing from the brain through

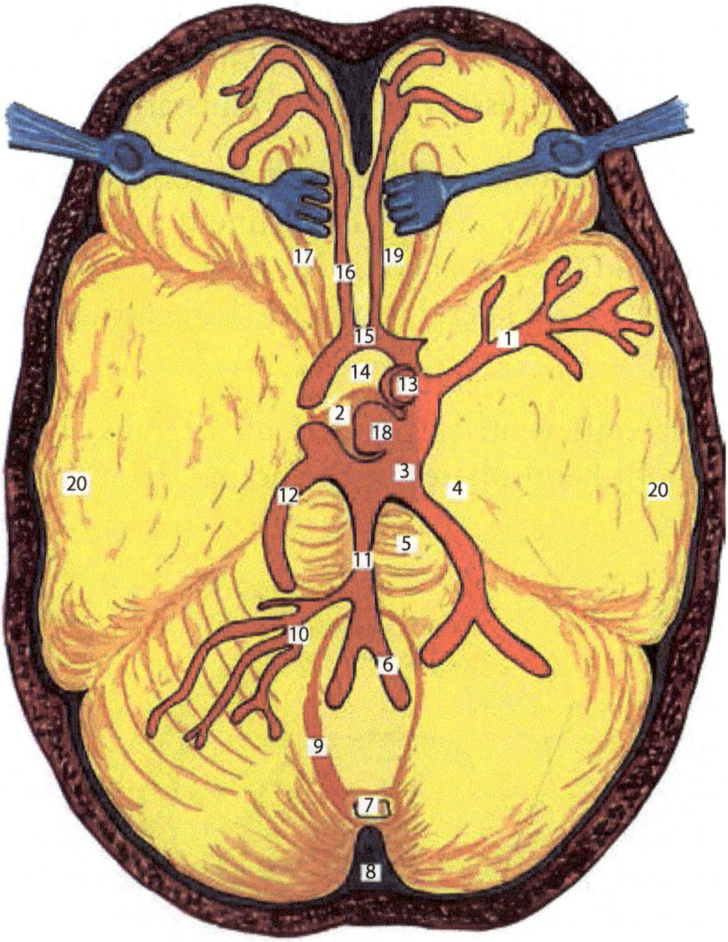

Figure 12 Cerebral vessels.

1 – a. cerebri media; 2 – infundibulum; 3 – a. commrtnicans posterior; 4 –pedunculi cerebri; 5 – pons; 6 – a. vertebralis; 7 – medulla oblongata; 8 – sinus occipitalis; 9 – a. cerebelli interior posterior; 10 – a.cerebelh inferior anterior; 11 – a. basilaris; 12 – a. cerebri posterior; 13 – a. carotis interna; 14 – chiasma opticum; 15 – a. communicans anterior; 16 – a. cerebri anterior; 17 – tr. olfactorius; 18 – aHeepmua a.commimicans posterior sinistra; 19 – fades medialis; 20 – fades lateralis the veins enters the sinuses of dura (Fig. 13-16). Dural sinuses communicate with the veins of soft tissues of the head and diploic veins using emissary veins (vv. Emissariae).

Liquor system of the brain. This system consists of the system of the ventricles and the subarachnoid space belonging to the brain and the spinal cord. The ventricular system has four ventricles - two lateral ventricles, III ventricle and and IV ventricle.

The side (lateral) ventricles (vertriculi lateralis) lie in the depths of both hemispheres of the brain. There are the left

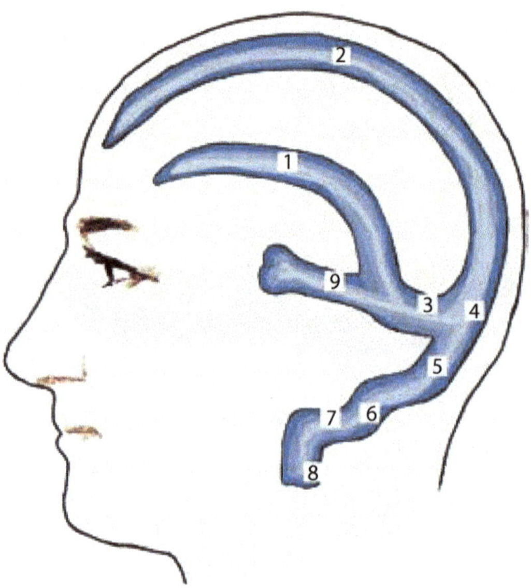

Figure 13 Dural venous sinuses

1 – sinus sagitalis inferior; 2 – sinus sagitalis superior; 3 – sinus rectus; 4 – torcular herofili; 5 – sinus transverses; 6 – sinus sigmoideus; 7 – bulbus venae jugularis; 8 – v. jugularis interna; 9 – v. cerebri magna

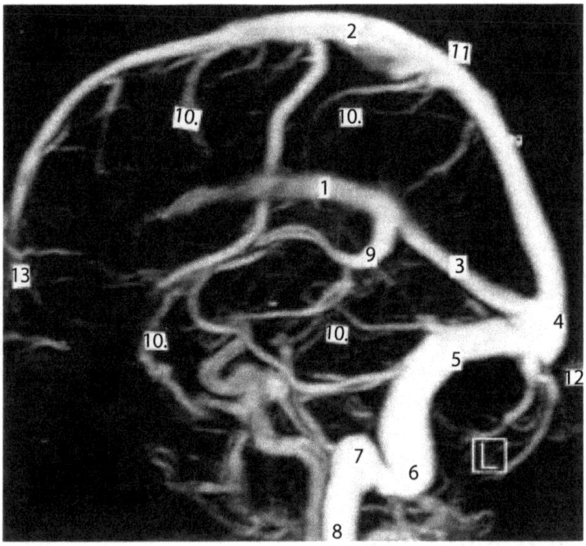

Figure 14 Dural venous sinuses. MRI.

1 – sinus sagitalis inferior; 2 – sinus sagitalis superior: 3 – sinus rectus: 4 – torcular herofili; 5 – sinus transversus: 6 – sinus sigmoideus; 7 – bulbus venae jugularis; 8 – v. jugularis interna; 9 – v. cerebri magna; 10 – вснм ranoBHoro M03ra; 11 – fissure saggitalis superior; 12 – protuberantia occipitalis interior; 13 – anastomoses with veins of nasal cavity

lateral ventricle (ventriculus lateralis sinister) and right lateral ventricle (ventriculus lateralis dexter), both which are placed in the respective hemispheres. Both lateral ventriculi have an anterior horn (cornu anterius), the central part (pars centralis), posterior horn (cornu posterius) and lower horn (cornu inferius). Third ventricle (ventriculus tertius cerebri) is unpaired. Its fissure-like cavity is located in the median sagittal plane and is connected with the lateral ventricles through the interventricular hole, and with the IV ventricle - through cerebral aqueduct (aqueductus cerebri). The fourth brain ventricle (ventriculus quartus cerebri) is also unpaired.

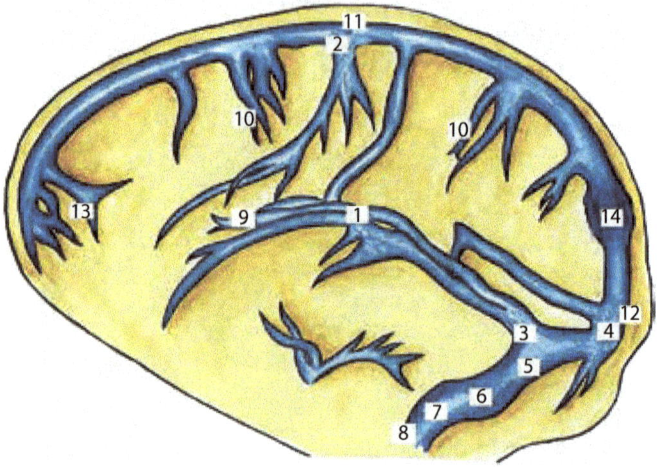

Figure 15 Dural venous sinuses

1 – sinus sagittalis inferior; 2 – sinus sagittalis superior; 3 – sinus rectus; 4 – torcular herofili; 5 – sinus transverses; 6 – sinus sigmoideus, 7 - bulbus venae jugularis; 10 – vv. encephali; 11 – sulcus sagittalis superior; 12 – protuberantia occipitalis interna; 13 – anastomoses with veins of nasal cavity; 14 – thrombosis of superior sagittal sinus.

It is surrounded with the pons and with the medulla oblongata from the front and with the cerebellum from the back and the sides. The rear portion of the fourth ventricle has two side apertures (apertura lateralis ventriculi quarti), or Luschka's holes, through which the ventricular cavity gets connected with the subarachnoid space. Choroid plexus of the ventricular system are the main source of cerebrospinal fluid (70-85%). Vascular plexi are literally the folds of the pia mater. They are well vascularized.

Cerebrospinal fluid constantly flows from the lateral ventricles through the interventricular foramen to the III ventricle, then continues through the cerebral aqueduct to the IV

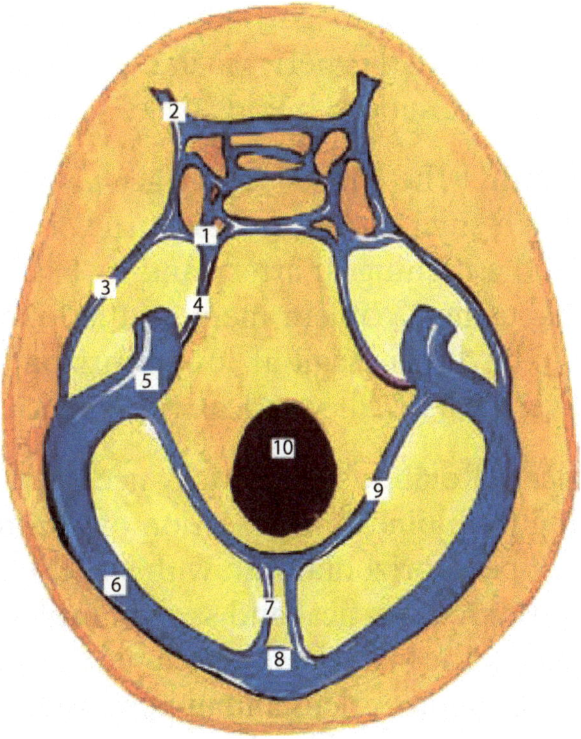

Figure 16 Venous sinuses of the dura mater

1 – sinus cavernosus; 2 – v.ophtalmica superior; 3 – sinus petrosus superior; 4 – sinus petrosus inferior; 5 – bulbus superior v. jugularis internae; 6 – sinus transversus; 7 – sinus occipitalis; 8 – confluens sinuum; 9 – v. cerebri magna; 10 – foramen magnum

ventricle. It departs through the middle and lateral foramens fluid from IV ventricle, and after that it enters the large tank, where it washes basal and convexital surfaces of the cerebral hemispheres and then falls into the subarachnoid space of the spinal cord. After that it returns to the subarachnoid space of the brain. The backflow of liquid from it is performed using filtration into the venous system - into sinuses of the dura mater, where blood comes through the arachnoid granulations. Backflow also partially goes on through the lymphatic system and the perivascular perineural fissures that connect with the subarachnoid space.

Brain Topography. The head skin surface projection of the gyri and the main sulci of the cerebral cortex, the brain ventricles, the middle meningeal artery and its branches, dural venous sinuses: using Krönlein method, the following lines are supposed to be drawn: sagittal, lower horizontal, top horizontal, front vertical, middle vertical, rear vertical.

In order to locate Rolando's sulcus it is necessary to draw a line connecting the point of intersection of the first vertical line and and upper horizontal line with the upper point of intersection of the rear vertical and sagittal lines. Sylvian fissure is projected by the bisecting of the angle composed of the projection line Rolando's sulcus and upper horizontal line. A. meningea media is projected by the intersection of the frontal vertical line with the upper edge of the zygomatic arch; its anterior branch can be found at the intersection point of the frontal horizontal line and upper vertical line, and the posterior branch - at the intersection of the rear vertical line with the same Krönlein line.

Sinus sagittalis is projected on the sagittal line, sinus transversus - across lin. nuchae superior, sinus sigmoideus – by the lower rear quadrant of the mastoid process.

The lateral ventricles are projected within the semi-circle with the radius equal to half of the distance between the external auditory canal and the sagittal line; in this case the external auditory canal is considered to be a center. The fourth ventricle is arranged in such a way so its short diagonal line falls a little bit below and in parallel to lin. nuchae superior, while the long one is located on the sagittal line drawn as an extension till rear edge of the foramen magnum.

Surgical Anatomy of Congenital Disorders

Brain hernias are development of the specific type disorders of the skull and the brain characterized by a defect in the frontal or occipital bone which serves as hernia gate and lets brain matter fall through it. The following types of brain hernias exist: meningocele, encephalocele and entsefalotsistotsele.

1. Meningocele happens when arachnoid and soft tunicas pass through a hole in the cranial bones.
2. Encefalocele also has brain tissue itself.
3. Encefalocystocele holds a part of the brain ventricle as well.

Hydrocephalus is the dilatation of the ventricular system of the brain and subarachnoid spaces caused by excessive amount of cerebrospinal fluid. Congenital form is characterized by the increase in the circumference of the skull at birth, alteration of body/head ratio, noticeable increase of the size of suturas, and larger fontanelles. The forehead is higher than usual, overhanging, facial skeleton is relatively smaller, eyelids are half-closed. Skin is thinned, stretched, quite hydropic. Skull circumference can reach up to 80-100 cm.

Pathotopography of the Cerebral Part of the Head

Hemorrhage in the brain due to aneurysm rupture (Fig. 17).

Subarachnoid hemorrhage is a sudden bleeding to the subarachnoid space. The most frequent cause of spontaneous bleeding is the ruptures of aneurysms. Hemorrhage occurs as

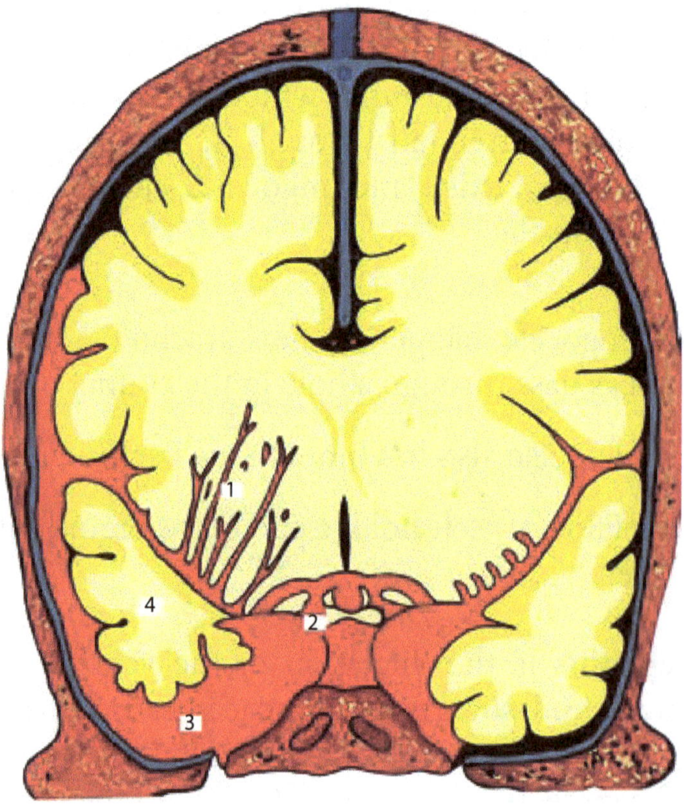

Figure 17 Bleeding in the brain due to rupture of the aneurism

1 - rupture of the artery; 2 - ruptured aneurysm (the circle of Willis); 3 - subarachnoid hemorrhage (blood comes from a torn aneurysm); 4 - compression of the temporal lobe blood coming from a torn aneurysm

a result of the release of blood from the ruptured aneurysm into the space between the arachnoid and the soft meninges. The most common cause is a traumatic brain injury, but traumatic subarachnoid hemorrhage is considered as an independent nosology. Spontaneous (primary) subarachnoid hemorrhage in about 85% of cases is due to rupture of intracranial aneurysms, most often congenital saccular or grozdepodobnyh. Bleeding can stop spontaneously. Aneurysm rupture can occur at any age, but more often occurs at the age of 40-60 years. Less frequent causes are mycotic aneurysms, arteriovenous malformations and diseases with hemorrhagic syndrome.

Pathotopography: with subarachnoid hemorrhage in the brain, there is a rupture of the arterioles, rupture of the aneurysm of the vessels of the Willis circle, subarachnoid hemorrhage (blood comes from the ruptured aneurysm), squeezing the temporal lobe with blood coming from the ruptured aneurysm. Intracranial hemorrhages. In young children, intracranial hemorrhages account for 56-60% of acute disorders of cerebral circulation. The causes of intracranial hemorrhages are ruptures of aneurysms, craniocerebral trauma, blood diseases (leukemia, anemia, thrombocytopenia and other disorders of the coagulation system), hemorrhagic vasculitis, brain tumors. Aneurysm rupture in most cases is accompanied by subarachnoid hemorrhage, while for blood diseases, asphyxia of newborns are more typical parenchymal small-point, perivascular hemorrhages. Massive hemorrhages in the brain substance and intraventricular hemorrhages are usually the result of rupture of an aneurysm or severe craniocerebral trauma and, as a rule, are not compatible with life. Types of intracranial hemorrhages: epidural - usually as a result of damage to the branch of the middle artery of the medulla; Subdural - from a damaged vein or cerebral sinus; Subarachnoid - due to damage to the soft shell of the brain. (Fig. 18).

Pathotopography: epidural hemorrhage is a blood clotting between the dura mater and the bones of the skull. The signs are: peeling of the dura from the bones of the skull, filling the epidural space with blood; Compression of the hard,

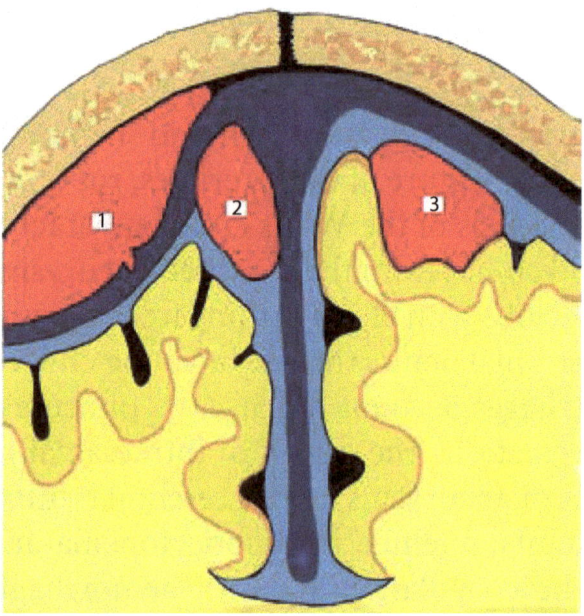

Figure 18 Types of intracranial hemorrhage

Epidural (1) is usually a result of damage to the branch of the middle cerebral artery membrane; subdural (2) of the damaged vein or cerebral sinus; subarachnoid (3) as a result of damage to the soft lining of the brain (M.Grob)

arachnoid and soft shells of the brain, as well as brain structures. With computed tomography, lenticular hyper-density volumetric formation adjacent to the bones of the cranial vault is revealed.

Subdural hemorrhage is a hemorrhage arising between the dura mater and soft mater. With this type of hemorrhage, blood filling of the subdural space is observed, compression of the arachnoid, soft membranes of the brain, compression of the brain structures. Computed tomography and magnetic resonance imaging are the most informative methods for diagnosing a subdural hemorrhage of hemispheric localization. Hemorrhage is visualized as a "sickle" zone of increased density, adjacent to the cranial vault.

Subarachnoid hemorrhage is a hemorrhage into the subarachnoid space (the cavity between the arachnoid and the soft medulla). The main signs of this type of hemorrhage: the filling of the subarachnoid space with blood, the compression of the arachnoid membrane and the structures of the brain. The first day is preferable for magnetic resonance imaging, since the blood can already be seen. At a later date, up to 3-5 days, computed tomography is preferable.

Hematomas of the brain. One of the types of brain injuries that occur when there is a blow to the head or in the case of a penetrating wound, in which there is a violation of its functions, is the brain's hematoma. Such traumatic injuries can be small, moderate or severe. Severe injuries can lead to loss of consciousness or death.

Hematoma refers to the accumulation of a limited amount of coagulated or liquid blood in a formed cavity with closed or open injuries of organs and tissues, with the wound of the vessels. Hematoma can be located in any part of the brain.

Types of hematomas of the brain: epidural, intracerebral (Fig. 19)

Pathotopography. ***Epidural hematomas*** represent a traumatic accumulation of blood, located between the inner surface of the bones of the skull and the dura mater, causing local and general compression of the brain. The frequency of epidural hematomas among the observations of primary hospitalized victims with craniocerebral trauma varies widely from 0.54% to 9%. Symptoms: epidural hematoma is characterized by the fact that its central part is thicker (2-4 cm) than the peripheral parts. Representing an incompressible mass consisting of liquid blood and its clots, epidural hematoma suppresses the underlying cerebral membrane and brain substance, forming a dent according to its shape and size. On computed tomography of the brain, the epidural hematoma looks like a biconvex lens.

Intracerebral hematoma is limited by the accumulation of blood in the brain substance. It squeezes, displaces and damages nearby brain tissue, up to its necrosis. On a computer tomography the intracerebral hematoma has the appearance of a center of homogeneous density of round or oval shape. If the hematoma is formed as a result of a brain contusion, then it usually has an uneven contour. With the

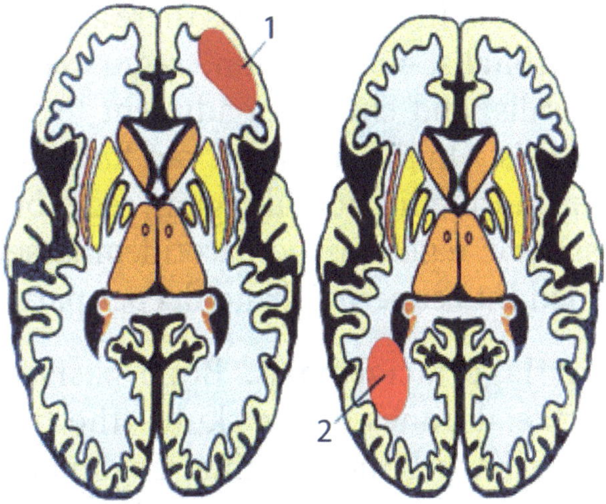

Figure 19 Different types of brain hematomas

Epidural hematomas are caused by the accumulation of blood injury that sits between the inner surface of the skull bones and dura mater, causing local and general compression of the brain. The frequency of epidural hematomas among primary observations of hospitalized patients with head injury varies widely from 0.54% to 9%.

same time, the density of the hematoma decreases to an iso-dense state.

Facial Head Region

Topography of the facial part of the head (......)

Boundaries. The border between the brain and the facial part of the head follows by the upper edge of the eye socket, through the fronto-zygomatic suture across zygomatic arch to the outer ear canal opening and to the top of the mastoid process.

The region's division. In the front area there are the eye socket region, regio orbitalis, nose region, regio nasalis, mouth region, regio oralis, and chin area adjacent to the previous one, regio mentalis. On the sides you can find the suborbital region, regio infraorbitalis, buccal region, regio buccalis, and parotideomasseteric region, regio parotideomasseterica. The former is divided into superficial and deeper parts.

Bone and cartilage facial basis. Bone basis is made of tightly joined facial bones of the skull - the frontal bone, os frontale; maxillary bone, maxilla; zygomatic bone, os zygomaticum; nasal bones, ossa nasalia and lower jaw, mandibula.

Cartilaginous base is formed by cartilages of the nose: the lateral cartilage of the nose - paired, the triangular cartilage – unpaired; large alar cartilage - unpaired, small nose cartilages - paired - three pieces on each side, they lie behind the larger cartilage of the nose ala. Sometimes it's possible to find there several additional nasal cartilage of different sizes between the lateral cartilage and a large cartilage of ala nasi. The edge of the unpaired cartilage of the nasal septum touches the inner surface of the nasal bridge. Between the lower edge of the cartilage of the nasal septum and the front edge of the vomer there is a narrow strip of the vomeronasal cartilage.

Individual and age differences. Individual differences in the form of external nose in each case depend on the characteristics of its bone and cartilage bases. Nose silhouette depends on the shape of the nasal bones and the angle between them and the frontal bone. Depth of the nose bridge is also a major factor. Radix nasi can be located deeper, shallower, higher or lower comparatively to the forehead. Nasal bridge can be straight, convex or depressed. Nose cartilages also play an important role. Depending on the angle of convergence of lateral cartilage, it can have different degree of flatness. Ala nasi's cartilages form the lower part of the nose. Depending on their vertical size of the ala can be longer or shorter. Nose points forward if the cartilages are long and massive. "Snub-nosed" nose is a result of a significant upward deviations of ala cartilages. "Hanged up" nose which sometimes completely covers the upper lip forms when ala cartilages tilt straight downwards.

The area of the orbit. Orbit, or orbita, is a paired symmetrical cavity of the skull. Eyeball with its supporting apparatus lies within its boundaries (Fig. 19-20).

Humans orbits have a form of square truncated pyramids. Their apexes are pointed back, to the sella Turcica in the cranial cavity, while their base is pointed to the front, to the surface. Axises of orbital pyramid converge in the rear direction and diverge in the forward direction. The average size of the orbit: the depth of the adult orbit ranges from 4 to 5 cm; width at the entrance is about 4 cm and the height is typically less than 3.5-3.75 cm.

Orbital walls are formed by various lamellae. They separate the eye sockets from the rest of the skull: the upper wall of the orbit goes from the anterior cranial fossa and the frontal sinus; the lower wall of the orbit proceeds from the maxillary sinuses; the medial wall of the orbit runs from the nasal cavity and the lateral wall proceeds from the temporal fossa.

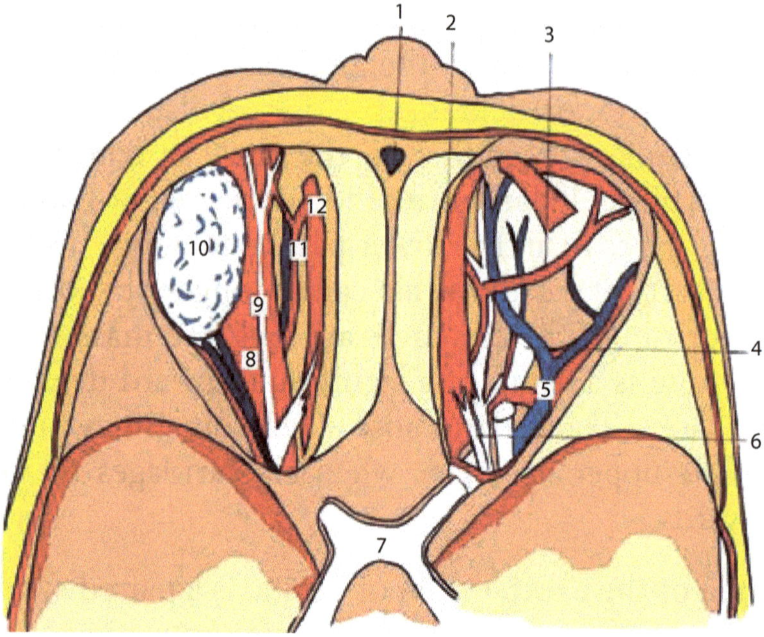

Figure 20 Orbital muscles, vessels and nerves. Topdown view,
1 – sinus sagittalis superior; 2 – m. obliquus superior; 3 – a. supraorbitalis; 4 n. opticus; 5 – v. ophtalmica superior; 6 – nn. trochlearis et oculomotorius; 7 – chiasma opticum; 8 – m. rectus superior; 9 – n. ophtalmicus; 10 – gl. lacrimahs; 11 – a. ophtalmica; 12 – m. rectus medialis

The area of the orbit has two regions: superficial is located in front of the orbital septum and forming regio palpebra, and profound is located behind orbital septum and forming regio orbitalis propria, in which the eyeball is laid with its muscles, nerves, fatty tissue and vessels.

There are anterior and posterior ethmoid foramina in the medial wall. Fissura orbitalis superior (located in posterior regio) connects orbit with fossa cranii superior. Fissure

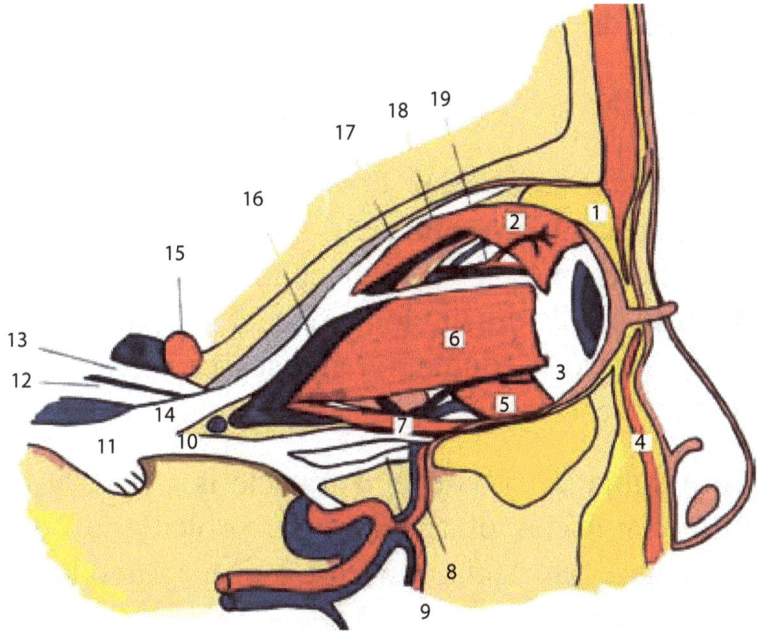

Figure 21 Orbital muscles, vessels and nerves. Side view.

1 – corpus adiposum orbitae; 2 – m. levator palpebrae superioris; 3 – bulbus oculi; 4 – m. orbicularis oculi; 5 – m. obliquus inferior; 6 – m. rectus lateralis; 7 – m. rectus inferior; 8 – a. infraorbitalis; 9 – n. infraorbitalis; 10 – n. maxillaris; 11 – gangl. trigeminale; 12 – n. trochlearis; 13 – n. oculomotorius; 14 – n. ophtalmicus; 15 – a. carotis interna; 16 – v. ovhtalmica suverior: 17 – n. frontalis: 18 – m. rectus superior: 19 – a. lacrimalis

orbitalis inferior between lateral and inferior walls connects orbit with temporal and infratemporal fossae, sphenoid sinus. The optic nerve has the central position in orbit.

Eyelids. Eyelids are plates made of skin and cartilages that are curved according to the shape of forward eye segment. They protect the surface of the eye.

Layers of eyelids:

- Skin of the eyelids is thin and movable.
- Subcutaneous cellular tissue of eyelids is loose, there are a lot anastomoses of the vessels of the eyeball and the vessels of the face inside it. That's why it easily swells during local (e.g. stye), and general (angioneurotic edema, renal diseases, etc.) inflammatory processes.
- Thin subcutaneous eyelid muscle is a part of the mimic muscles of the eye, m. orbicularis oculi, and, like other facial muscles, is innervated by the facial nerve.
- There is a layer below the muscle. This layer is composed of the eyelid cartilages and the orbital septum attached to them. The septum is attached to the supra-and the infraorbital edges.

The posterior surface of the cartilage century and orbital septum is lined with mucous membranes - the conjunctiva, or conjunctiva palpebrarum, which goes on into the sclera of the eyeball, conjunctiva bulbi. Places where conjunctiva transits from eyelids to sclra form the upper and lower conjunctiva's domes - fornix conjunctivae superior et inferior. The lower dome can be examined by pulling down the lid. For the inspection of the upper conjunctiva's dome it is necessary to turn the upper eyelid out.

The frontal edge of the eyelids has eyelashes, which have sebaceous glands at their basements. Purulent inflammation of the glands is known as stye - cholasion. It is possible to see the holes of specific sebaceous, or meibomian, glands landed in the thickness of the eyelid cartilage closer to the rear edge of the eyelids. The movable edges of the eyelids of the medial and lateral angles of the eyes form angles that are fixed to the bones of the eye socket with tendons.

Eyeball. It is located in the cavity of the eye socket, even though holding it only partially. Eyeball is surrounded by fascia, eyeball vagina, or vagina bulbi, or Tenon's capsule. Tenon's capsule, which covers the eyeball almost across its entire length, except for the region corresponding to the cornea (in the front) and the place of passage of the optic nerve (in the rear), suspends the eyeball in the orbit among the fat cellular tissue, while being fixed with the fascial strands that go to the wall of the eyesocket and its edges. Capsule walls are being pierced with the tendons of muscles of the eyeball. Tenon's capsule doesn't stick tightly with the eyeball: there is always a small fissure, spatium episclerale, which allows the eyeball to move easily.

Postbulbar part is located behind Tenon's capsule. Muscular apparatus of eyesockets includes 6 extraocular muscles (4 rectus muscles and 2 oblique muscles) and the muscle that lifts the upper eyelid (m. Levator palpebrae superior). Lateral rectus muscle is innervated by n. abducens, top oblique - by n. trochlearis, and the others, including the muscle that lifts the upper eyelid, are innervated by n. oculomotorius.

Postbulbar part. It has adipous tissue, ligaments, muscles, blood vessels and nerves. It has the following boundries: sclera limits it from the front; optic canal orbital aperture limits it from behind. Length of postbulbar part is about 25-35 mm.

Nasal region. The upper boundary of the nasal region is the horizontal line connecting the medial ends of the eyebrows (the root of the nose), the lower boundary is the line drawn through the attachment of the nasal septum and the lateral boundaries of the nasal area are defined by the nasobuccal and nasolabial folds. The nasal region is divided into the outer nose and nasal cavity.

Osteo-cartilaginous base. External nose, nasus externus, is formed with the nasal bones at the top, while at the sides it is formed with the frontal process of the maxilla and cartilages. The side wall of the nose is the lateral cartilages of the nose. Triangular cartilage is located below the nasal bones. At the lower point lateral cartilage joins on each side with the larger ala nasi cartilage. At the same time it is attached to the lower end of the nasal bone and the frontal bone of the upper jaw from behind. Greater ala nasi cartilage is paired and is located below the corresponding lateral cartilage of the nose, limiting entrance to the nasal cavity. Smaller ala nasi cartilages are paired. There are three of them on each side. They lie behind the larger ala nasi cartilage. Sometimes you can find additional cartilage of variable sizes between the lateral cartilage and a larger ala nasi cartilage. Internally, by the nasal cavity, cartilages of the nasal septum lie adjacently with the inner surface of nasal bridge. Nasal septum cartilage is unpaired, has 4-angled polygon shape and forms a large frontal part of the nasal septum. In the rear and above the cartilage of the nasal septum connects with the perpendicular plate of the ethmoid bone, and in the rear and below it does so with the vomer and the frontal nasal spine. Between the lower edge of the cartilage of the nasal septum and the front edge of the vomer there is a narrow strip of the vomeronasal cartilage situated on each side.

Nasal cavity (Fig.22). Cavum nasi is the primary division of the respiratory tract. It contains within itself olfactory organ. Apertura piriformis nasi comes in in its front, and paired holes, choanae, connect it to the nasopharynx from behind. Nasal cavity is divided into two not quite symmetrical halves with the bone of the nasal septum, septum nasi osseum. Each half of the nasal cavity has five walls: top wall, bottom wall, rear wall, medial wall and lateral wall.

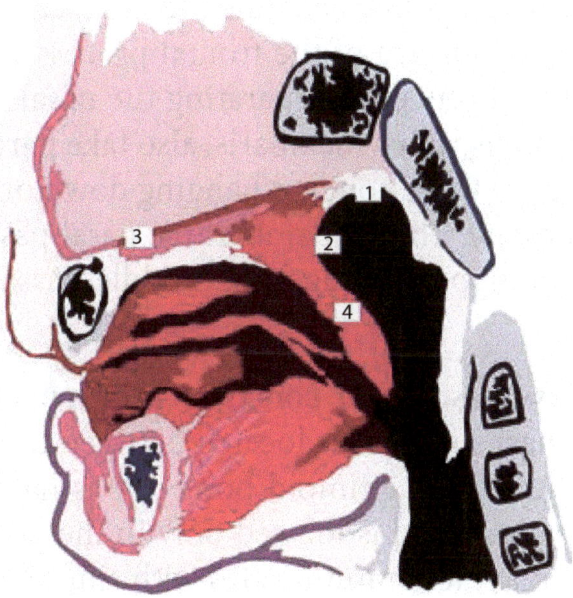

Figure 22 The nasal cavity

1 - paries superior; 2 - ostium pharyngeum tubae auditivae; 3 - palatum durum; 4 - palatum molle

The upper wall of the nasal cavity is formed by a small part of the frontal bone, lamina cribrosa of the ethmoid bone and part of sphenoid bone. The bottom wall of the nasal cavity, or bottom, includes palatine process of the maxilla and the horizontal plate of the palatine bone that together form up the hard palate, palatum osseum. The bottom of the nasal cavity serves as a "roof" for the oral cavity. Nasal septum is the medial wall of the nasal cavity. The rear wall of the nasal cavity goes only to a small extent and is present only in the upper section since otherwise it would block hoanas lying below. It is formed by the nasal surface of the body of the sphenoid bone with the twin foramens present on it – apertura sinus sphenoidalis. Lateral wall of the nasal cavity is formed by the lacrimal bone, os lacrimale, and lamina orbitalis of the ethmoid bone which together separate the nasal cavity from the eye socket. Nasal surface of the frontal process of the upper jaw and the thin bony plate separating the nasal cavity from the maxillary sinus, sinus maxillaris, also take part in the formation. There are three conches hanging down on the lateral wall of nasal cavity, They separate three nasal passages from each other: the upper passage, the middle passage and the lower passage.

The nasal passages. Nasal conches. The upper nasal passage, meatus nasi superior, is located between the upper and middle conches of the ethmoid bone; the is half as long as the average passage and is located only in the posterior part of the nasal cavity; it communicates with sinus sphenoidalis, foramen sphenopalatinum. Rear cells of the ethmoid bone are also opened inside. Middle nasal passage, meatus nasi medius, goes between the middle and lower conches. Cellulae ethmoidales anteriores et mediae and sinus maxillaris are also opened inside. The lower nasal passage, meatus nasi inferior, passes between the lower conch and the bottom of the nasal cavity. The nasolacrimal duct opens in its lower part. The space between the conches and nasal septum is marked as a common nasal passage. On the side wall of the nasopharynx there is a pharyngeal opening of the auditory tube that connects the pharyngeal cavity with the middle ear (tympanic cavity). It is located at the rear end of the lower conch about 1 cm behind it.

The vessels of the nasal cavity form the anastomotic nets which are created by multiple systems. The arteries relate to branches of a. ophthalmica (aa. ethmoidales anterior and posterior), a. maxillaris (a. sphenopalatina) and a. facialis (rr. septi nasi). Vein form nets which tend to be found more superficially. Really dense venous plexus (which look like cavernous formations) are swarmed under the mucosal tissue of the lower and middle conches. Most nosebleeds originate from there. Veins of nasal cavity make anastomoses with the veins of the nasopharynx, orbits and the brain tunicas.

Sensory innervation of the nasal mucosa is being done through the 1st and the 2nd branches of the trigeminal nerve, that is, optical and maxillary nerve. The specific innervation is carried on by the olfactory nerve.

Paranasal sinuses. There are maxillary and frontal sinus, ethmoid labyrinth and partly sphenoid sinus present on each side of the nasal cavity.

Highmore sinus, or sinus maxillaris, is located inside of the maxillary bone. Its is the largest sinus among all the paranasal sinuses; its capacity in adults varies from 10 to 12 cm3 (Fig.

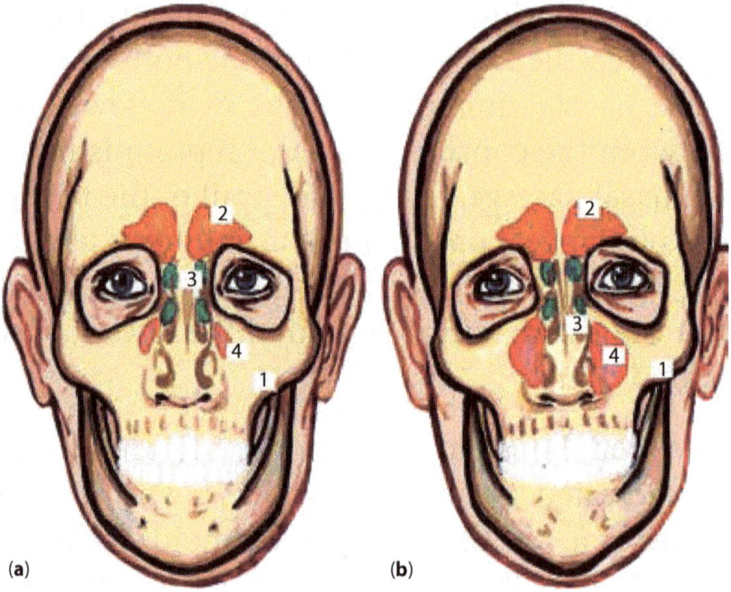

Figure 23 Variants of pneumatization of maxillar sinuses depending of maxilla structure: 1 - maxilla; 2 - sinus frontales; 3 - cellulae labyrinthus ethmoidalis; 4 - sinus maxillaris

23). The shape of the maxillary sinus resembles the tetrahedral pyramid whose base lies on the lateral wall of the nasal cavity, and the tip points at the zygomatic process of the maxilla.

The anterior wall faces forward, top, or orbital, sinus wall separates the sinus from the eye sockets, and the posterior wall faces the infratemporal and pterygopalatine-palatal fossae. The bottom wall of the maxillary sinus is formed by the alveolar bone of the upper jaw, which separates sinus from the oral cavity.

The inner, or nasal, sinus wall corresponds to the biggest part of the lower and middle nasal passages. The foramen, through which the maxillary sinus communicates with the nasal cavity, hiatus maxillaris, is located right beneath the bottom of the orbit. Such location contributes to the stagnation of the liquids excreted during sinus inflammations. The nasolacrimal duct goes across the anterior part of the inner wall of the sinus maxillaris, while celluli ethmoidali attach to the upper posterior part. The upper or orbital wall of the maxillary sinus is the thinnest one, especially in the posterior part. Inside the orbital wall threre is a canal of the infraorbital nerve. Sometimes nerves and blood vessels lie directly with the sinus' mucosa. The frontal or the facial wall of the maxillary sinus of the upper jaw is formed by the part located between the infraorbital margin and alveolar process. It is the thickest wall of the maxillary sinus; it is covered with soft cheeck tissue and it can be easily palpated. The flat imprint in the center of the front surface of the anterior wall, which is called "fossa canina", is the thinnest part of this wall. At the upper edge of the "fossa canina" threre is a foramen of the infraorbital nerve, foramen infraorbitale. Rr. alveolares superiores anteriores et medius (branch n. infraorbitalis II branch of trigeminal nerve), pass through this wall, forming a plexus dentalis superior. Aa. alveolares superiores anteriores also pass through this wall from suborbital artery (from a. maxillaris). The bottom wall, or bottom of the maxillary sinus, is located near the posterior part of the alveolar bone of the upper jaw and usually corresponds to the fossae of four upper posterior teeth. This makes it possible to open the maxillary sinus.

Frontal sinus, *sinus frontalis,* is located between the plates of orbital part and the squama of the frontal bone. It is divided into lower or orbital, anterior or front, back or brain, and the median walls.

Frontal sinus communicates with the nasal cavity through the apertura sinus frontalis, which opens in front part of the middle nasal meatus.

Sphenoid sinus, *or sinus sphenoidalis,* is located within the body of the sphenoid bone directly behind the ethmoid labyrinth above choanas and fornix pharyngis. The sinus with the sagital septum is divided into two unequal (in most cases) parts. On the front side of the thinnest wall there is an opening, apertura sinus sphenoidalis, which can be found in each half of the sinus. The shape and size of the sphenoid sinus are highly variable. Its upper wall is turned to the front and the middle cranial fossae.

The middle part of the upper wall of the sphenoid sinus corresponds to the sella Turcica with hypophisis located in its fossa.

Optic chiasm can be found slightly forward. Internal carotid artery and the cavernous venous sinus go outside the wall of the sphenoid sinus. In addition, in each sinus side you can find oculomotor, trochlear and abducens nerves, perforating the outer wall of the cavernous sinus, as well as the Ist branch of the trigeminal nerve. The bottom wall of the sphenoid sinus forms the body of the nasal cavity.

Ethmoid labyrinth, *labyrinthus ethmoidalis,* consists of 2-5 sometimes more ethmoid cells of different size and shape, cellulae ethmoidales, which are separated from the anterior cranial fossa with the orbital part of the frontal bone and the ethmoid bone, from the orbit – with the orbital plate, lamina orbitalis. The sphenoid plate of ethmoid labyrinth is the free edge which serves as the basis of the middle conch, and divides the air cells into front cells and posterior cells. The front cells open into the middle nasal meatus and the posterior ones – into the upper meatus. The mucosa of the paranasal sinuses is slightly different from the nasal mucosa in its structure, but it is much thinner and has less vessels and glands than the nasal mucosa. Blood supply is performed via branches of internal and external carotid arteries, mainly through the ophthalmic and maxillary arteries.

Veins of the Haimore's sinus anastomose with the facial veins and the pterygoid plexus, while the veins of the frontal sinus – with the veins of the dura mater, the longitudinal sinus and with the cavernous sinus. Sometimes an infection might get into cranial cavity or eye-socket through these pathways.

The innervation of the paranasal sinuses is carried out by I and II branches of the trigeminal nerve, as well as by the ganglion pterigopalatinum.

Mouth area is an oral cavity and lip region.

Oral cavity divides into oral cavity and mouth vestibulum.

Vestibule of Mouth, *vestibulum oris* is a fissure-like horse-shoe-shaped space which communicates with the environment through oral slit – rima oris. Formix vestibuli oris is a part of vestibule of the mouth where lips and cheeks mucosa extends to the mucosa of alveolar parts of the jaws. The mucous membrane passes from lips to gums and forms transitional folds.

Oral cavity. When the mouth is closed its cavity is a narrow horizontal fissure formed by the arch of the hard palatinum and tongue; the side edges of the tongue touch the jaws and lingual surfaces of the teeth.

Boundaries. The anterolateral wall of the oral cavity is formed by the alveolar processes with teeth and partialy by the body and inner surface of the mandible branches and the medial pterygoid muscles. At the posterior part the oral cavity opens at the isthmus faucium and goes into the middle part of the pharynx, pars oralis pharyngis. At the top this part is connected with the nasal cavity through the nasal pharynx and hoanas, choanae. At the bottom it is linked with the cavity of the larynx and esophagus though the hypopharynx, pars laryngea pharyngis.

Soft palate, *palatum molle,* is the posterior wall of the oral cavity. It consists of symmetrically arranged muscles of the soft palate and tongue muscles. The upper wall of the oral cavity is formed by the hard palate. Near the front end of the longitudinal palatal suture, almost directly near the cervices of the central incisors, you can find incisal opening, foramen incisivum which leads into the channel with the same name. N. nasopalatinus from the II branch of the trigeminal nerve passes through it. In the posterolateral corners of the palatinum threre are symmetrically arranged large and small palatal holes, foramina palatina majores et minores, belonging to canalis palatinus major. This canal connects pterygopalatine fossa with oral cavity and serves as the passage of the palatine nerves, nn. palatini anteriores, mediales, and posteriores, and the descending palatine artery, a. palatina descendens. The throat opening is formed by the contraction of the soft palate muscles between its edge, the front and back of the tongue arches.

The posterior edge of the soft palate transforms into the side wall of the fauces in the shape of two folds of the anterior and posterior palatine arches. In the anterior arch there is the palatoglossus muscle, m. palatoglossus, and in the posterior arch there is m. palatopharyngeus. Fossa tonsillaris can be found between the arches where tonsils, tonsilla palatinae, are located. The tonsilla palatinae is separated from the pharyngeal wall by with loose connective tissue.

Palatine tonsil has its own capsule, capsula tonsillae, and is covered with mucosa. The blood supply of tonsils is provided by ascending pharyngeal and facial arteries (branches of the external carotid artery), and also by the descending palatine artery (from a. Maxillaris). Tonsils nerves are the branches of the glossopharyngeal nerve (IX pair), n. vagus (X pair), the lingual nerve (coming from the 3rd branch of the trigeminal nerve) and ganglion pterygopalatinum. They come to the amygdala from the outside. Palatal tonsils together with the pharyngeal tonsils (on the rear wall of the nasopharynx), lingual tonsil (which is located behind the base of the tongue) and two pipe tonsils (they can be found at pharyngeal opening of auditory (eustachian) tube) form the pharyngeal lymphoid ring, anulus lymphoideus pharynges. Primarily it was described by Nikolai Pirogov, and then by Waldeyer.

The innervation of mucosa of the hard and soft palate is carried by the II branches of the trigeminal nerve through the ganglion pterygopalatinum, which gives beginning to the palatine nerves, nn. palatini. The muscles of the soft palate are innervated in the following way: the muscles which pull the soft palate, m. tensor veli palatini, are supplied with IIIrd branch of the trigeminal nerve; other muscles are innervated by the branches of the pharyngeal plexus.

The upper and lower jaw. Both jaws are similar in general anatomical structure: each has three arcs: dental arc, alveolar arc and basal arc.

The upper jaw consists of two fused bones. The middle part of it is called the intermaxillary bone. It is thin, has air inside, and contains the Haimore's sinus. This bone is adjacent to the orbital and nasal cavities. It is fused with the facial bones, which is why it is immovable. The upper jaw has almost no attachment points of the masticatory muscles. A few fibers of the external pterygoid muscle are attached to the region of tuberositas of the upper jaw.

The lower jaw bone is unpaired. Its middle part is occupied with symphysis. During the first year of life both halves grow into one bone as a result of synostosing. It is made of a compact bone. It is quite firm and contains a narrow mandibular canal. It is the only facial bone which can move and attached to the base of the skull. The whole inner surface of the mandible as well as its processes serve as an attachment area for masticatory muscles. The lower jaw has a horseshoe shape. It is divided into body, alveolar process, and two branches; each branch ends rising with the two processes: the frontal process with coronal (proc. coronoideus) and the posterior one with articular (proc. condylaris), the upper part of which is called the articular head. There is Incisura mandibulae between the processes.

Dentes – Teeth

The anatomical and clinical formula of deciduous and permanent teeth. Teeth are classified into non-permanent, primary teeth which fall off during early childhood and permanent teeth. A child usually has 20 teeth of the first category. On each side of the child's mouth there are two incisors, two canines, two molars and no premolars. The medial incisors erupt at 6-8 months, lateral incisors - at 7-8 months, front molars - at 12-15 months, canines - at 15-20 months, posterior molars - at 20-24 months. There are 32 permanent teeth: there are two incisors and three premolars on each side. Sometimes eighth teeth, so-called "wisdom teeth", do not erupt at all. Each tooth has the crown (corona dentis) covered with enamel and the root of the tooth - radix dentis. Radix grows from tooth alveola, alveola dentis. Tooth cervix and radix are covered by cement, cementum. Cement is surrounded with periodontium. Tooth pulp, pulpa dentis is located inside tooth corona. It gradually transforms into tooth cervix and radix, and after that into the tooth radix canal, canalis radios dentis. Blood supply of the teeth: the teeth of the upper jaw are supplied with a. maxillaris, posterior top teeth - with the posterior upper alveolar artery, middle and front teeth of the upper jaw are supplied with the branches of the infraorbital artery, a. infraorbitalis, - upper front and middle alveolar arteries, aa. alveolares superiores anteriores et mediae. The teeth of the lower jaw are supplied with blood from the inferior alveolar artery, a. alveolaris inferior. It is a branch of a. maxillaris. The innervation of the teeth: the upper jaw is innervated via n. maxitlaris, the lower jaw - via n. mandibulans. The correlation between the upper row of teeth and the lower row of teeth is called articulation or bite. In the correct bite upper dental arch extends slightly forward above the lower dental arch.

***Tongue,** or lingua, is divided* into three parts: radix, corpus and apex.

The basis of the tongue consists of the skeletal muscle and its own ones:

- M. genioglossus - genioglossus muscle starts from spina mentalis of mandible. In the contraction of this muscle tongue leans out of the oral cavity.

- M. hyogtossus – sublingual muschle. It starts at the large horns of the hyoid bone. In contraction this muscle retracts the tongue into the oral cavity.

- M. styloglossus – styloglossus muscle. It starts from processus styloideus, and pulls the tongue backward and upward;

- M. palatoglossus - palato-lingual muscle is located at the palatal arc with the same name.

The proper tongue muscles are m. longitudinals linguae - longitudinal muscle, m. transversus linguae - transverse muscle, m. verticalis linguae - vertical muscle. Tongue is supplied through a. lingualis, a branch of a. carotis externa. The innervation of the tongue is performed by n. lingualis - lingual nerve, a branch of n. mandibularis, chorda tympani -tympani, a branch of n. facialis – they are responsible for the taste feelings of the frontal two-thirds of the tongue; n. glossopharyngeus - glossopharyngeal nerve is responsible for the taste feelings of the back one-third of the tongue; n. laryngeus superior - upper laryngeal nerve is responsible for the innervation of the posterior part of the tongue; n. hypoglossus - hypoglossal nerve – it provides motor innervation of the tongue. Lymph backflow: submandibular lymph nodes collect lymph from frontal parts of the tongue, retropharyngeal lymph nodes collect lymph from posterior parts.

Sublingual salivary glands. Gl. sublingualis are located just under the mouth mucosa, inside the sublingual space near the mylohyoid muscle, next to the excretory duct of the submandibular gland. Sublingual salivary glands are the smallest ones among the "big" salivary glands. A single gland weights about 5 g. Glands have flattened oval shape. The main excretory duct opens into the sublingual caruncle. Small ducts open into the sublingual fold. Blood supply is carried out through branches of the lingual artery.

The fundus of the oral cavity. It is formed by the complex of soft tissues which are located between the tongue and the skin of suprahyoid part of anterior neck region. Mouth fundus is composed of oral diaphragm, diaphragma oris, formed by the geniohyoid muscle, m. myiohyoideus. Above the diaphragm in the sagittal direction there is m. geniohyoideus - geniohyoid muscle. Tongue – lingua – can be found right above it. Between the radix and the corpus of the tongue there is a blind foramen, foramen caecum. During the embryonic period it opens into the ductus thyreoglossus.

Buccal region (Fig. 24). External marks of the buccal region are the zygomatic bone and zygomatic arch, the bottom edge of the lower jaw, nasolabial fissure, and the anterior edge of m. masseter. The limits of the buccal region: the upper limit of the buccal region is zygomatic arch, the lower limit is the lower edge of the lower jaw, the anterior limit is the vertical line drawn from the outer corner of the eye, and the posterior limit is the palpable anterior edge of the masticatory muscle.

In the buccal region the cellular tissue is the most developed and Bichat's fat pad (located between the buccal and masticatory

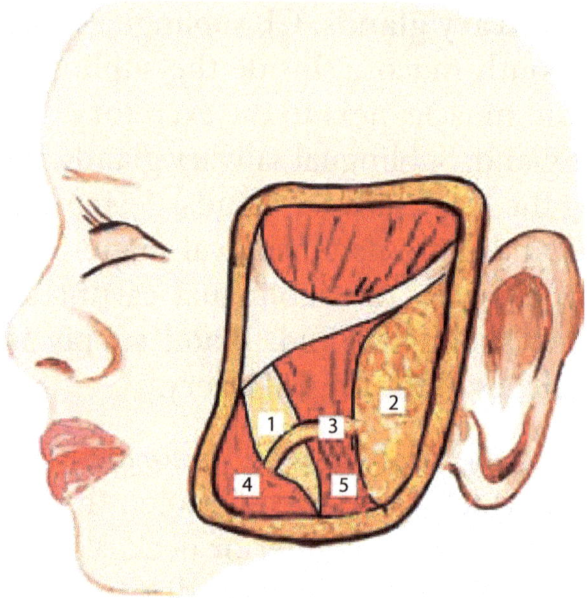

Figure 24 Buccal region.

1 – corpus adiposum buccae; 2 – gl. parotis; 3 – ductus parotideus; 4 – m. buccinators; 5 – m. masseter

muscules) is attached to it. Muscles of expression of the buccal region are presented by the lower part of m. orbitalis oculi, m. quadratus labii superiors, m. zygomaticus. Afferent nerves are the branches of n. trigeminus – n. infraorbitalis and nn. buccalis. Efferent nerves are the branches of n. facialis.

Layers and their characteristics. The skin of the buccal region is quite thin and contains a large amount of sweat and sebaceous glands. It is firmly attached with the well-developed layer of the cellular tissue. Facial artery and vein proceed inside the cellular tissue.

Buccal fat pads are practically important formations which are located in the subcutaneous layer. They are also known as Bichat's fat pads and corpus adiposumbuccae. They are located at the back of the posterior limit and attach to the frontal edge of the masticatory muscle. Buccal fat pads are enclosed into a quite tight fascial capsule, which separates it from the cellular tissue and buccal muscle located deeper.

One part of the pad is located in an adjacent, parotid-masticatory area, between the deep surface of the m. masseter and the m. buccinator. There are several processes coming from this part of the pad: temporal, orbital and pterygo-palatine, that proceed into the corresponding regions. The temporal process follows the zygomatic bone along the outer wall of the orbit through masticatory-maxilar space and reaches the front edge of the temporal muscle. Here this process attaches with subfascial temporal space and the deeper temporal space (between the bone itself and deep surface of temporal muscle). Orbital pad process is located in the infratemporal fossa adjacent to the lower orbital fissure. Pterygoid-palatal pad process proceeds further into the outer basis of the cranium between the posterior edges of the upper and lower jaw and the base of the pterygoid process. Sometimes this process reaches the lower medial part of the superior orbital fissure and enters into the the skull cavity through it, where it attaches to the wall of intra-cavernous dural sinus.

Blood vessels. The blood supply is performed by a. carotis externa: a. temporalis superficialis, a. ficialis, maxilaris and a. ophthalmica (from a. carotis interna). Facial artery and vein in the buccal region are projected from the intersection of the anterior edge of the masseter muscle with the lower edge of the lower jaw in the diagonal direction to the inner corner of the eye. One of the most important facial vein anastomoses with pterygoid plexus can be found on this line approximately on the level of ala nasi. There are two venous nets on the face: superficial (consists of facial and submandibular veins) and profound (is presented by pterygoid venous plexus).

The projection of the branches of the facial nerve, parotid gland, and the exit points of trigeminal nerve's branches. Motor branches of the facial nerve that innervates the mimic muscles are projected along the lines diverging in a fan manner from a point lower and forward to the tragus. The nerves go through bone foramena. Foramen infraorbitale is projected 0,5 cm below the middle of infraorbital margin. Foramen mentale is projected on the middle body of the mandibular between 1 and 2 premolars. Foramen mandibular is projected in the middle poit between the anterior and posterior edge of the mandibular branch for 2,5-3 cm upwards from the lower edge.

To memorize the direction of facial nerve branches is easier in the following way: place the hand with the separated fingers so that the Ist finger would vertically cross the middle zygomatic arch, the IInd finger would go to the outer corner of the eye, the IIIrd - above the upper lip, the IVth – along the edge of the mandible, and the Vth – would be directed vertically downwards to the neck. When you place your hand in this way rr. temporales will match the Ist finger; rr. zygomatici – the IInd, rr. buccales – the IIIrd, r. marginalis mandibularis – the IVth and r. colli – the Vth finger. Excretory duct of the parotid gland, or Stensen's duct, is projected along the line parallel to the zygomatic arch and below it for 1.5-2.5 cm in the direction of the external auditory canal to the middle distance between ala nasi and the mouth corner.

Profound face area (Fig. 25-27). *Blood supply.* Maxillary artery (a. maxillaris) begins at the cervix of the lower jaw and then extends horizontally. It supplies blood to the deeper parts of the face. Maxillary artery is divided into three segments:

1. The first segment (mandibular part) is located medially to the branches of the mandible. The following arteries begin from the first segment:

    - Deeper ear artery (a. auricularis profunda), which supplies the joint capsule of the temporomandibular joint, as well as the lower wall of auditory canal and the eardrum.

    - Anterior tympanic artery (a. tympanica anterior), which supplies the tympanum.

    - Inferior alveolar artery (a. alveolaris inferior), which supplies the teeth. Its terminal branch is the mental artery (a. mentalis). It goes through foramen with the same name (foramen mentale) and then appears in the chin area.

    - The middle meningeal artery (a. meningea media) goes through the spinous foramen (foramen spinosum). After that it enters the cranial cavity and divides into anterior and posterior branches (ramus anterior et ramus posterior). The artery carries blood to dura mater.

    - Additional meningeal branch (ramus meningeus accessorius) proceeds through the foramen ovale (foramen ovale) and enters the

Figure 25 Profound face area.

1 – m. masseter; 2 – a. masseterica; 3 – a. carotis externa; 4 – substantia spongiosa mandibulae; 5 – n. auriculotemporalis; 6 – collum mandibulae; 7 – m. temporalis; 8 – a. et n. temporalis profunda; 9 – a. buccalis; 10 – m. pterygoideus externus; 11 – r. marginalis mandibulae; 12 – m. pterygoideus internus

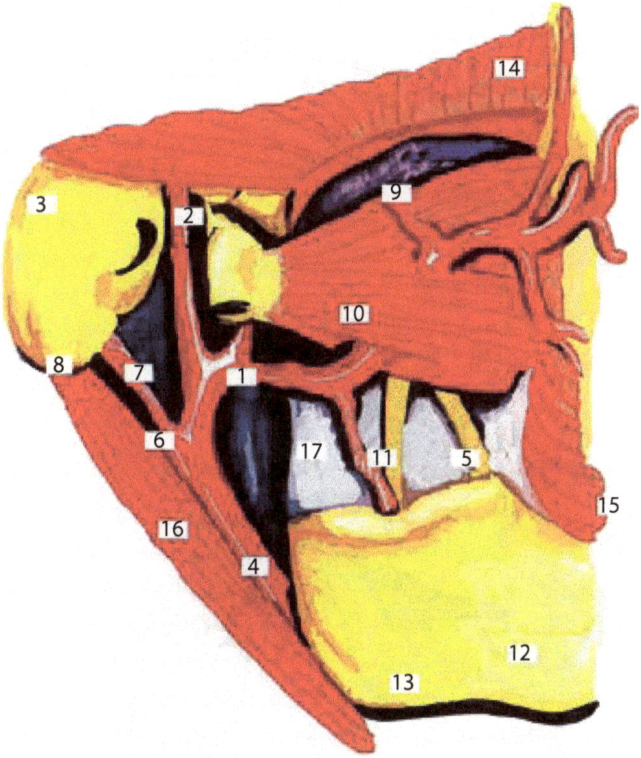

Figure 26 Profound face area.

1 - a. maxillaris; 2 - a. temporalis superficialis; 3 - porus acusticus externus; 4 - a. carotis externa; 5 - n. linguales; 6 - a. occipitalis; 7 - a. auricularis posterior; 8 - processus mastoideus; 9 - a. meningea media; 10 - m. pterygoideus lateralis; 11 - n. alveolaris inferior; 12 - corpus mandibulae; 13 - basis mandibulae; 14 - n. temporalis; 15 - porus acusticus externus; 16 - m. buccinator; 17 - m. digastricus (venter posterior); 18 - a. alvealaris inferior

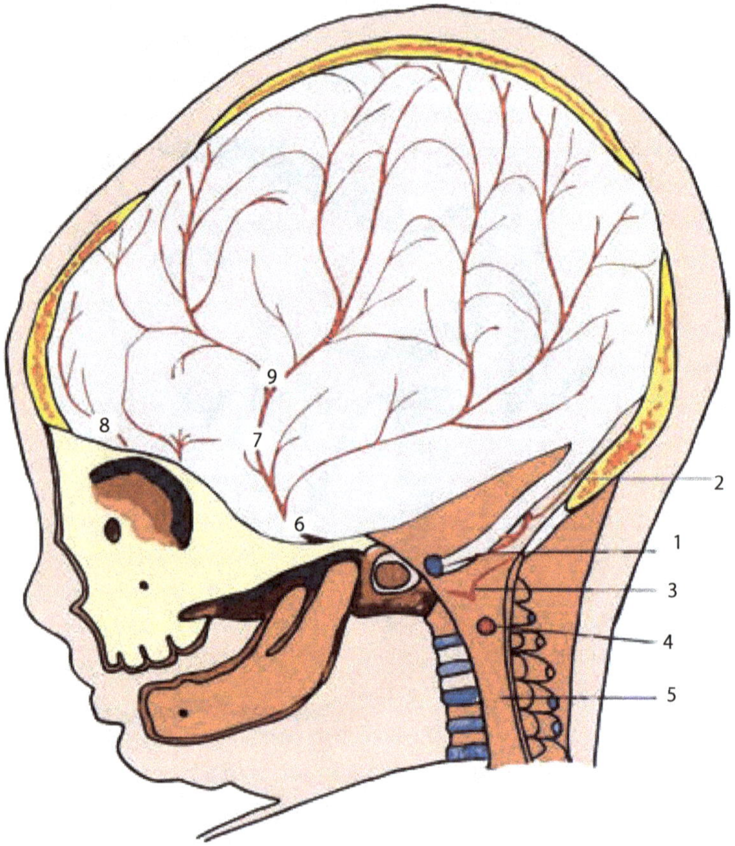

Figure 27 Dural arteries. Side view

1 - ramus parietalis a. meningeae medie; 2 - sinus sigmoideus; 3 - r. meningeus posterior a. vertebralis; 4 - a. vertebralis; 5 - medulla spinalis; 6 - a. meningea media; 7 - r. frontalis a. meningea mediae; 8 - a. meningea anterior; 9 - rr. terminales a. meningea mediae

cranial cavity. There it carries blood to the trigeminal ganglion (ganglion trigeminale).

2. The second section (pterygoid section) is located in the temporopterygoid space (spatium temporopterygoideum). It has the following branches:

- artery of masseter muscle (a. masseterica) which makes loop across the mandibular notch (incisura mandibulae) and enters the masseter muscle (m. masseter).

- pterygoid branches (rr. pterygoidei) supply blood to the outer and inner pterygoid muscles.

- deeper temporal arteries (aa. temporales profundae) supply blood to the temporal muscle.

- buccal artery (a. buccalis) is directed onwards and supplies blood to the muscles of the region of the same name.

- posterior upper alveolar artery (a. alveolaris superior posterior) enters with its terminal branches through the alveolar foramens (foramina alveolaria) into alveolar process of the upper jaw and supplies blood to the superior posterior teeth.

3. The third segment (pterygopalatine part) corresponds to the fossa pterygopalatina. It has the following branches:

   - descending palatine artery (a. palatina descendens) goes downward by the large palatine canal; after that it bends down, and enters the large palatal foramen (foramen palatinum majus) located on the hard palate and takes the horizontal direction along the alveolar process of the upper jaw;

- sphenoid-palatine artery (a. sphenopalatina) through the foramen with the same name enters the nasal cavity, and supplies its posterior part with the blood;

- suborbital artery (a. infraorbitalis) comes out of the fossa pterygoidea and through the lower orbital fissure (fissura orbitalis inferior) enters the eye socket, then proceeds inside the infraorbital sulcus (sulcus infraorbitalis), then – inside the infraorbital canal (canalis infraorbitalis). It leaves through the infraorbital foramen (foramen infraorbitale) and splits into multiple branches within the canine fossa (fossa canina).

Blood outflow. Within infratemporal and pterygopalatine fossae there is pterygoid venous plexus (plexus pterygoideus), which accepts the blood coming from the following vessels:

- middle meningeal veins (vv. meningeae mediae)

- deeper temporal veins (vv. temporales profundae)

- parotid veins (vv. parotidei)

- anterior ear veins (vv. auriculares anteriores)

- stylomastoid vein (v. stylomastoidea) Pterygoid plexus makes anastomoses with:

- the facial vein (v. faciallis) with the help of the profound face vein (v. faciei profunda)

- the retromandibular vein (v. retromandibularis) through the maxillary veins (vv. maxillares)
- the cavernous sinus by the venous plexus of foramen ovale (plexus venosus foraminis ovalis) and venous plexus of carotic canal (plexus venosus caroticus internus)
- the upper and lower ophthalmic veins (vv. ophthalmici superior et inferior)

Innervation. Mandibular nerve (n. mandibularis) is multipurposed; it has sensory fibers coming from the trigeminal ganglion (ganglion trigeminale) and fibers of motor radix (radix motoria). On leaving cranial cavity through foramen ovale it devides into the following sensory branches:

- Meningeal branch (ramus meningeus) passes through the spinous foramen (foramen spinosum) to the dura mater.
- Auriculotemporal nerve (n. auriculotemporalis) penetrates the parotid gland and innervates the posterior parts of the temporal region.
- Buccal nerve (n. buccalis) innervates the buccal skin and mucosa.
- Inferior alveolar nerve (n. alveolaris inferior) enters the mandibular foramen (foramen mandibulare), where it forms the lower dental plexus (plexus dentalis inferior). The terminal branch of the inferior alveolar nerve is the mental nerve

(n. mentalis), which branches into smaller nerves in the chin area and supplies the skin of the chin and lower lip after leaving through the mental foramen (foramen mentale).

- Lingual nerve (n. lingualis). It connects with the chorda tympani's taste fibers which serve as an innervation of anterior two thirds of the tongue.

The branches of the mandibular nerve give sensory fibers coming to the parasympathetic ganglions: ear ganglion (ganglion oticum), submandibular ganglion (ganglion submandibulare) and sublingual ganglion (ganglion sublinguale).

Cellular spaces of profound face areas (Fig. 28). They include temporomandibulopterygoid and interpterygoid cellular spaces and retromandibular, pterygoid fossae.

Temporomandibulopterygoid space (spatium temporomandibutopterygoideum) is located between the lower jaw, the temporal muscle and the ptrerygoid muscle located deeper.

Interpterygoid space (spatium interpterygoideum) is locked between the two pterygoid muscles. Maxillary artery and its branches, the venous pterygoid plexus, plexus venosus pterygoideus, are also located here. By the course of the lingual nerve this space communicates with the adipose tissue of the mouth. That is why the phlegmons developed here can spread to adipose tissue of the oral cavity bottom. Pus can also get into pterygoid fossa if it flows forward.

Retromandibular fossa is a depression located behind the ascending branch of the mandible. It has the following limits: anterior limit is a branch of the mandible, ramus mandibulae,

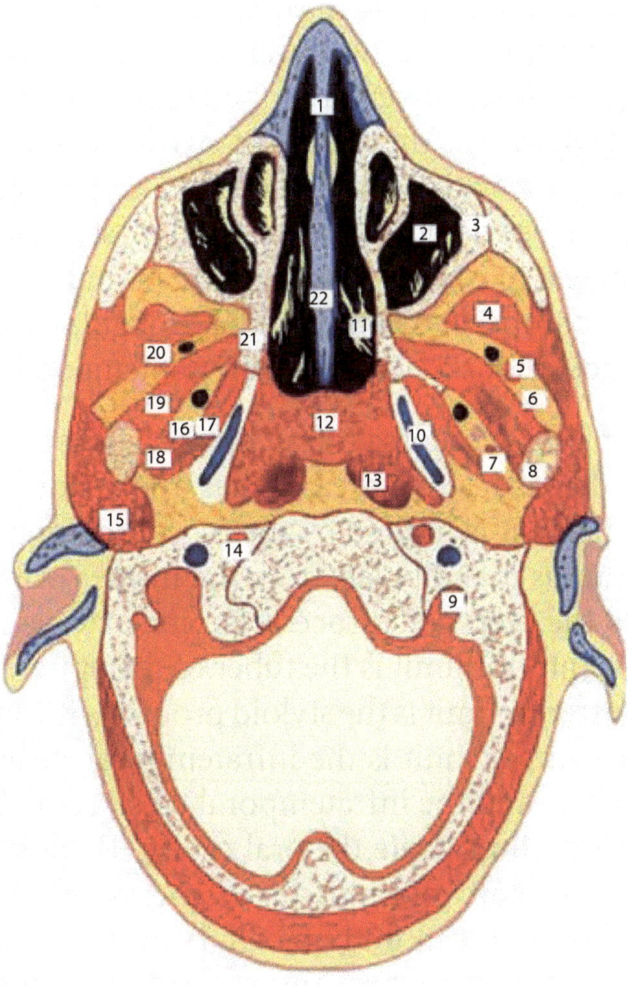

Figure 28 Cellular spaces of profound face areas

1 – septum nasi; 2 – sinus maxillaris; 3 – arcus zygomaticus; 4 – m. temporalis; 5 – a. mandibularis; 6 – n. lingualis; 7 – a. mandibularis; 8 – processus articularis mandibulae; 9 – incisura jugularis ossis occipital; 10 – pars cartilaginosus tubae auditivae; 11 – choana; 12 – tunica mucosa faucium et choanae; 13 – glandulae tunicae submucosae faucium; 14 – a. carotis interna; 15 – gl. parotis; 16 – n. alveolaris inferior; 17 – m. pterygoideus medialis; 18 – interpterygoid cellular space; 19 – m. pterygoideus lateralis; 20 – temporopterygoid cellular space; 21 – basis processus pterygoideus ossis basillaris; 22 – septum nasi

posterior limit is a mastoid process, processus mastoideus, upper limit is the outer ear meatus, meatus acusticus externus, bottom limit – posterior venter of the digastric muscle, venter posterior m. Digastrici. The bottom of this depression is formed by styloid process with m. stylohyoideus, stylohyoid muscle, m. styloglossus – styloglossus muscle and m. stylopharyngeus, stylopharyngeus muscle. Within retromandibular fossa there are glandula parotis, a. carotis externa, v. jugularis externa, a. auricularis posterior, n. facialis, n. auriculotemporalis.

Infratemporal fossa or fossa infratemporalis is located deeper than parotid-masticatory area. It has the following limits: from the outside it is limited by the ascending branch of the mandible, ramus mandibulae, from the inside - with the outside plate of the pterygoid process, lamina externa processus pterygoidei; anterior limit is the tuber of the upper jaw, tuber maxillae; posterior limit is the styloid process with anatomical muscle heap; upper limit is the infratemporal surface, facies infratemporalis, and the infratemporal crest, crista infratemporalis; bottom limit is the the oral cavity. The fossa is filled with the lateral pterygoid muscle, m. pterygoideus lateralis. Inside the fossa you can find: maxillary artery, a. maxillaris, with the branches of its first and second segments. Behind the articular process there are 5 branches: A. auricularis profunda, a. tympanica, a. meningea media, a. alveolaris inferior, ramus meningeus accessories.

The following arteries are located at the level of incisura mandibulae and processus coronoideus: a. temporalis profunda anterior, a. temporalis profunda posterior, a. masseterica, a. buccalis, a, pterygoidea lateralis, a. pterygoidea medialis, a. alveolaris superior.

Veins of infratemporal fossa form pterygoid plexus, plexus pterygoideus. This plexus widely anastomoses with v. ophthalmica inferior, v. facialis and v.jugularis interna through fissura orbitalis inferior.

Infratemporal fossa communicates with pterygoid fossa - fossa pterygopalatina, which is limited with tuber of the upper jaw, tuber maxillae, from the front, and with the pterygoid process, processus pterygoideus from behind, with the vertical palatal plate - medially, and with the larger wing of the sphenoid bone from the top. Pterygoid fossa communicates with the orbit through the lower orbital fissure, fissura orbitalis inferior, and does so with the nasal cavity through the foramen sphenopalatinum, which is located on the medial wall of the pterygoid fossa. It also links with the mouth through canalis palatinus major, opens into smaller and larger palatine foramens, foramen palatinum major et minor. It also communicates with the the middle cranial fossa through a round foramen - foramen rotundum, and with the outer cranium base surface - through the pterygoid canal, canalis pterygoideus. Inside pterygoid fossa there is the terminal section of the jaw artery, from which within this fossa the following branches start: a. alveolaris superior posterior, a. infraorbitalis, a. palatina descendens, a. sphenopalatina. N. maxillaris, n. trigemini and gangl. sphenopalatinum are also present in this fossa.

Parotid-masticatory region (regio parotideomasseterica) has the following limits: top limit is the zygomatic arch (arcus zygomaticus), bottom limit is the bottom margin of the lower jaw (margo inferior mandibulae), anterior limit is the anterior margin of the masseter muscle (m. masseter), posterior limit is the retromandibular fossa.

This Area Has Such Layers:

1. Derma - or the skin.
2. Panniculus adiposus - subcutaneous adipose tissue.
3. Fascia superficialis - superficial fascia.

Fascia parotideomasseterica - parotid-masticatory fascia – covers the glands from all sides except its upper margin, and by giving connective tissue septums into the depth of the gland this fascia divides the gland into segments. Lobed structure plays a major role in migrating nature of gland inflammation. During purulent parotitis abscess is usually being drainaged through the external auditory canal because of the absence of the fascia on the upper margin of the gland and because gland lies closely to the external auditory canal, which is why pus can easily break out through the incisura cartilaginis meatus acustici.

After breaking out pus infiltrates parapharyngeal space, and from there across the pharynx and esophagus it continues to flow into posterior mediastinum which causes mediastenitis.

4. Glandula parotis is parotid gland. In its depth you can find:
   - A. carotis externa or external carotid artery is located deeply inside the gland and divided into two terminal branches - a. maxillaris and a. temporalis superficialis;
   - N. auriculotemporatis or auliculotemporal nerve accompanies a. temporatis superficialis;

- V. retromandibulans - retromandibular vein;
- N. facialis or the facial nerve pierces gland and falls into pieces, forming a large goose foot. Its branches create plexus parotideus in the depth of the gland;
- Ductus parotideus (Stenoni) is an excretory duct of the parotid gland, located in the horizontal way. When it reaches the front margin of the m. masseter it turns 90 degrees inside, pierces m. buccinator and goes backwards. That is why its shape resembles the letter "T". At the level of the sixth or seventh superior tooth there is a place where parotid duct opens into the vestibulum of the oral cavity. Stenon's duct is projected from the bottom of the earlobe to the corner of the mouth. M. buccinator is a sphincter of the parotid gland's duct. Usually the duct is closed, and the saliva doesn't enter the oral cavity. But in chewing and dilation of the buccal muscle the duct opens and the saliva flows freely into the oral cavity;
- Nodi lymphatici parotidei superficiales et profundi - superficial and profound parotid lymph nodes

5. M. masseter - masticatory muscle. It is located at the anterior region of the parotid-masticatory area.
6. Os mandibula or the lower jaw is located in the posterior part of parotid-masticatory area. The upper segment of the ascending branch of the

mandible has a notch - incisura mandibulae. That is the point where a. et, n. masseterica come into the masticatory muscle from the depth.

7. Temporomandibular joint (articulatio temporomandibularis) is located in the posterior superior part of the parotid-masticatory region.

Infraorbital region (Fig. 29-31) (regio infraorbitalis) is limited with the infraorbital margin (margo infraorbitalis) from above, and with the base of the upper lip at level of the upper fornix of mouth vestibulum from below, with zygomaticomaxillaris suture from the outside, and with ala nasi from the inside.

The Layers of Suborbital Region:

1. The skin (cutis)
2. Fat deposits (panniculus adiposus)
3. Superficial mimic muscles layer.
   - The lower part of the orbital section of the orbicular eye muscle (pars orbitalis m. orbicularis oculi)
   - The muscle lifting the upper lip and ala of the nose (m. levator labii superioris alaeque nasi)
   - Muscle lifting the upper lip (m. levator labii superioris)
   - Zygomaticus minor muscle (m. zygomaticus minor)

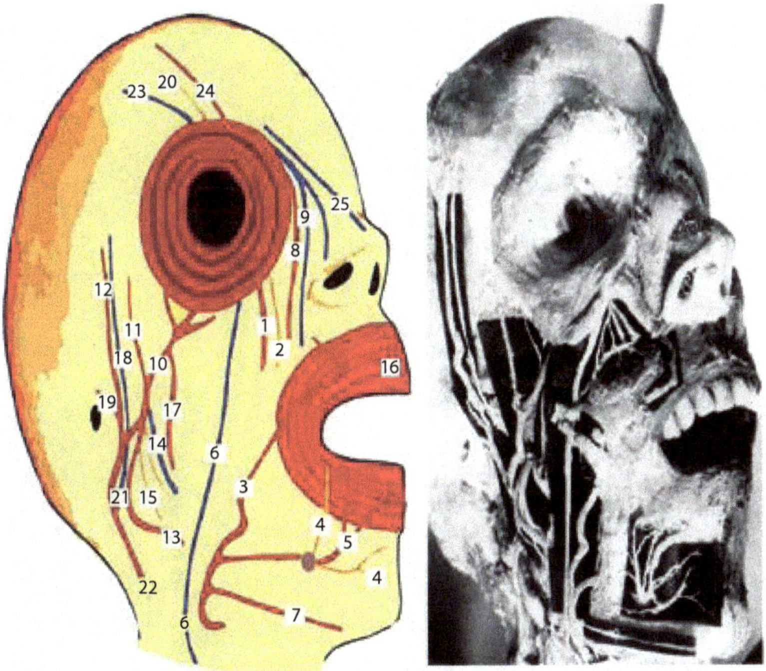

Figure 29 Regi o infraorbitalis

1 – a. infraorbitalis; 2 – n. infraorbitalis; 3 - a. facialis; 4 – n. mentalis; 5 - a. mentalis; 6 – v. facialis; 7 – a. submentalis; 8 – a. angularis; 9 – v. angularis; 10 – a. maxillaries; 11 – a. temporalis profunda; 12 - a. temporalis externa; 13 – a. alveolaris inferior posterior; 14 – v. alveolaris inferior; 15 – n. alveolaris inferior; 16 – a. labialis superior; 17 – a. alveolaris superior posterior; 18 – v. temporalis externa; 19 – r. auricularis anterior a. temporalis suprafacialis; 20 - r. supraorbital n. trigemini; 21 – v. maxillaris; 22 – a. carotis externa; 23 – v. supraorbitalis; 24 – a. supraorbitalis; 25 – v. nasalis externa.

Figure 30 Regio infraorbitalis

1 – v. temporalis superficialis; 2 – a. temporalis superficialis; 3 - a. angularis; 4 – a. maxillaris; 5 - a. temporalis profunda; 6 – r. buccalis a. maxillarii; 7 – a. alveolaris superior posterior; 8 – r. orbitalis a. maxillarii; 9 – a. alveolaris inferior; 10 – n. alveolaris inferior; 11 - v. alveolaris inferior; M 12 – v. retromandibularis; 13 – a. carotis communa; 14 – a. infraorbitalis; 15 – n. infraorbitalis.

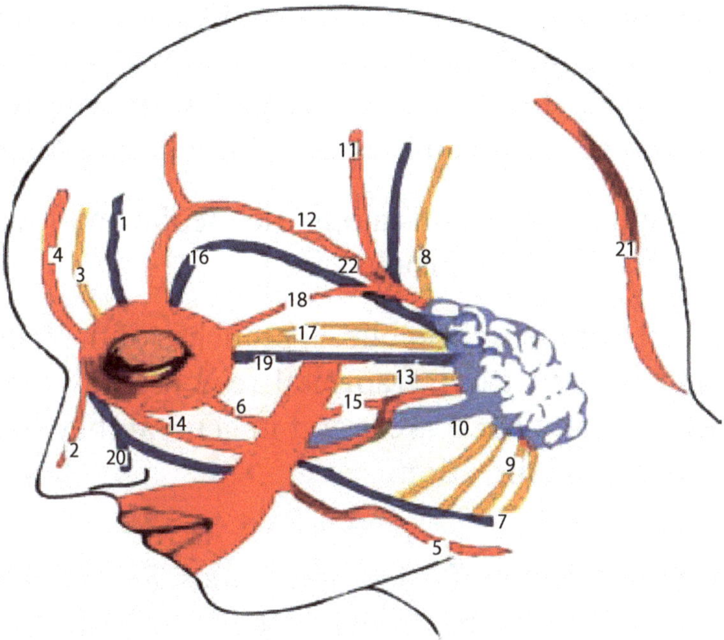

Figure 31 Infraorbital region

1 – v. supratrochlearis; 2 – a. dorsalis nasi; 3- r. supraorbitalis; 4 – a. supratrochlearis; 5 – a. facialis; 6 – a. infraorbitalis; 7 – v. facialis; 8 – n. auriculotemporal; 9 – plexus intraparotideus; 10 – gl. parotidea; 11 –a. temporalis superficialis; 12 –a. supraorbitalis; 13 –n. buccalis; 14 – a. infraorbitalis; 15 – a. maxillaris; 16 – v. supraorbitalis; 17 – n. opticus; 18 – a. temporalis profunda anterior; 19 – v. transversa faciei; 20 – v. nasals externae; 21 – a. occipitalis; 22 – r. frontalis a. temporalis superficialis.

4. Under the superficial layer of muscles there are angular artery and vein (a. et v. angularis), and also zygomatic and buccal branches of the facial nerve (rami zygomatici et buccales n. facialis).

- Angular vein is located in the medial corner of the eye and creates anastomoses with supratrochlear and supraorbital veins (vv. supratrochleares et supraorbitales), and also with the upper eye vein (v. ophthalmica superior). Angular vein is directed downward and laterally, crossing the infraorbital area diagonally at the same time.

- Angular artery runs along the medial and lower limits of the suborbital area.
- And zygomatic and buccal branches of the facial nerve enter the infraorbital region from zygomatic and buccal areas and in general they have horizontal direction.

5. Profound layer of the mimic muscles includes muscle that lifts the corner of the mouth (m. levator anguli oris), and buccal muscle (m. buccinator).
6. The anterior surface of the upper jaw (facies anterior maxillae) with the canine fossa (fossa canina) and suborbital foramen (foramen infraorbitale). Infraorbital vessels and nerve (a., v. et n. infraorbitales) enter the canine fossa through suborbital foramen. Suborbital vessels bring the blood to the soft tissues around canine fossa. The branches of the infraorbital nerve innervate the skin from the lower eyelid to the upper lip.

Chin area (regio mentalis) is separated from the lower lip with mento-labial fold (sulcus mento-labialis). From the bottom it is limited by the margin of the lower jaw, and from the outside by the vertical lines drawn down from the corners of the mouth.

The Chin Area Has the Following Layers:

1. The skin (cutis) is tightly joined with the musculofascial layer lying deeper. The skin itself is innervated by the mental nerve (n. mentalis).

2. Musculofascial layer (stratum musculofasciale). This layer has the following formations:

   - Mimic muscles: muscle depressing mouth angle (m. depressor anguli oris), transverse muscle of chin (m. transversus menti), muscle lowering inferior lip (m. depressor labii inferioris), chin muscle (m. mentalis).

   - Marginal mandibular branch of the facial nerve (ramus marginalis mandibulae n. facialis) innervates the mimic muscles.

   - Chin vessels and nerves (a., v. et n. mentalis) supply blood and provide sensory innervation of the area.

3. The periosteum.

4. The body of the mandible (corpus mandibulae) has the mental foramen (foramen mentale) at the level of the second molar, which is the entry point of blood vessels and nerves with the same name.

The Lymphatic System of the Head

Lymphatic vessels of soft tissues of the head and face are organized into the following groups of lymphatic nodes:

- Occipital lymphatic nodes (nodi lymphatici occipitales) accept lymph from the occipital region; they are located at the attachment point of the trapezius muscle (m. trapezius).
- Mastoid lymph nodes (nodi lymphatici mastoidei) lie behind the ear; they accept lymph from the outer ear and partially from the parietal and temporal regions.
- Superficial parotid lymph nodes (nodi lymphatici parotidei superficiales) are located in front of the tragus; they accept lymph from the frontal and partially from parietal and temporal areas.
- Deeper parotid lymph nodes (nodi lymphatici parotidei profundi) lie below the external auditory meatus; they accept lymph from the external auditory meatus, tympanic membrane and ear conch.
- Intraglandular profound parotid lymph nodes (nodi lymphatici parotidei profundi intraglandulares) lie in the depth of the parotid gland (glandula parotis), accompanying the external carotid artery; they carry lymph from the parotid gland and the tissues around it.
- Buccal node (nodus buccinatorius) is located in the buccal region after the masseter muscle (m. masseter) on the outer surface of the buccal muscle (m. Buccinator); it accepts lymph coming from the frontal face surface.

- Anterior submandibular lymph nodes (nodi lymphatici submandibulares anteriores) are located after the submandibular salivary gland; they carry lymph coming from the buccal nodes and the frontal face surface.
- Posterior submandibular lymph nodes (nodi lymphatici submandibulares posteriores) lie behind the submandibular gland; they carry lymph from the superficial parotid lymph nodes and adjacent face parts.
- Lower submandibular lymph nodes (nodi lymphatici submandibulares inferiores) are located below the submandibular gland; they take in lymph coming from the anterior and posterior submandibular lymph nodes.
- Upper submandibular lymph nodes (nodi lymphatici submandibulares superiores) are located above the submandibular salivary gland. They are not always present.
- Submental lymph nodes (nodi lymphatici submentales) lie in the chin triangle; they accept lymph coming from the chin area. There is usually one or two of them.
- Lingual lymph nodes (nodi lymphatici linguales) are located on the sides of the tongue radix along the side surface of the hyoglossus muscle (m. hyoglossus); they accept lymph radix along the side surface of the hyoglossus muscle (m. hyoglossus); they accept lymph coming from the tongue.

- Retropharyngeal deeper cervical lymph nodes (nodi lymphatici cervicales profundi retropharyngeales) can be found on either side of the posterior pharyngeal wall; they accept lymph from the posterior parts of the oral cavity, lymphoepithelial ring, pharyngeal walls and the middle ear.

So, the lymph flows from the head to the neck, where it passes through the superficial and profound cervical lymph nodes (nodi lymphatici cervicales superficiales et nodi lymphatici cervicales profundi). During this process, the superficial cervical lymph nodes receive lymph from the neck, mastoid, and superficial parotid lymph nodes; and the profound nodes receive lymph coming from the profound parotid, submandibular, submental and retropharyngeal lymph nodes.

Congenital Face Disorders

Macrostomy is a horizontal fissure. It is a congenital defect of the soft tissues of the mouth angle and cheeks, which results into extension of the mouth fissure. It can reach the ear accompanied by muscle hypoplasia, inability to close the mouth and drooling.

Coloboma is an oblique lateral face fissure, which stretches from the inner eye angle to the lower lip. In some cases, fissure affects only soft tissues and can be seen only near inner eye angle, but in the other ones a complete disintegration of all tissues with hypoplasia of the skeleton and with the invasion into the nasal cavity can be observed. Colobomas are often accompanied with conjunctivitis.

Fissure of the upper lip is one of the most common congenital disorders of the face. It is most commonly seen in boys and can be accompanied with a fissure in the palatinum. In case of incomplete fusion of upper lip defect affects only soft lip tissues. In case of complete infusion defects can be observed not only in the soft tissues, but also in the upper jaw. Deformation of nasal septum and flattening of the ala of the nose can also be seen on the affected side. In case of complete double-sided infusion intermaxillary process is shifted forward and held with the vomer, covered with small, isolated skin piece and the vermillion lip border. In this case breastfeeding is completely impossible.

Hard palatinum fissure appears as a result of the delay in the development of palatal processes of the upper jaw, which is why the latter does not reach the vomer and does not join with it. In the case of incomplete fusion it is possible to observe splitting of the tongue, splitting of both tongue and soft palate; splitting of the tongue, soft and hard palate. In case of complete infusion in addition to phenomena described above it is possible to observe the infusion of maxillar alveolar process. Complete palatinum infusion is usually combined with infusion of the upper lip. During one-sided palatinum infusion fissure connects the oral cavity with the one half of the nasal cavity. During two-sided palatinum infusion fissure connects the mouth with the both halves of the nasal cavity, so that vomer becomes visible in the middle of the fissure. Usually this disorder is accompanied by hyperplasia of maxilla and shortening of soft palatinum.

Pathotopography of Facial Part of the Head

Abscesses of the face (Fig. 32).

Abscess is a purelent inflammation of tissues with their melting and formation of the purelent cavity. It can develop in subcutaneous tissue, muscles, bones, organs and between them. The cavity of abscess can have both a simple round form and complex form with many pockets. At first walls of abscess are covered by purelent-fibrous impositions and pieces of necrotic tissues. Then the inflammation zone develops around the abscess periphery, that leads to the formation of a pyogenic membrane formed by the connective tissue.

Patotopography. Abscess of infratemporal fossa is a purelent-inflammatory process limited by the bounds of cellular space of this fossa.

As the oval and spinous foramena open in the upper wall of the fossa there is the possibility of the infectious-inflammatory process spreading into the brain coat. Infratemporal fossa contains the lower segment of temporal muscle and pterygoid muscles; maxillary, middle meningeal, lower alveolar, profound temporalis, buccal arteries and pterygoid venous plexus; mandibular, lower alveolar, lingual, buccal nerves, chorda tympani and ear ganglion. Due to the abscess occurrence there is a high pressure in this region, which leads to the ompression of the muscles, arteries and nerves in the infratemporal fossa. Profound temporal and buccal arteries shift to the lateral plate of the pterygoid process of the sphenoid bone. The buccal nerve shifts towards the zygomatic arch, lower mandibular and lower alveolar nerves - closer to the lower jaw.

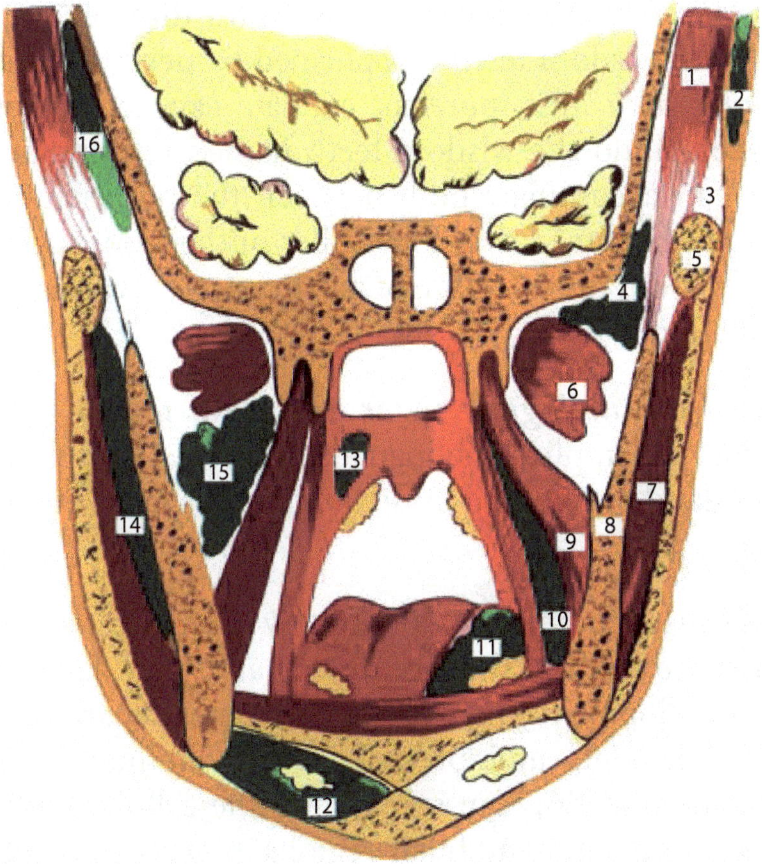

Figure 32 Localization of phlegmons

1 – m. temporalis; 2 – superficial phlegmon of the temporal region; 3 – aponeurosis temporalis; 4 – infratemporal fossa abscess; 5 – arcus zygomaticus; 6 – m. pterygoideus externus; 7 – m. masseter; 8 – r. ascendes mandibullae; 9 – m. pterygoideus internus; 10 – peripharyngeal space phlegmon; 11 – phlegmon of the mouth's bottom; 12 – submandibular space abscess; 13 – peritonsillar abscess; 14 – phlegmon chewing jaw-space; 15 – phlegmon of intrapterygoid space; 16 – deep phlegmon of the temporal region

Abscess of retromolar cavity. In case of the difficult eruption of the lower wisdom teeth complicated by pericoronitis and retromolar periostitis, purelent exudate can accumulate in the space behind the wisdom teeth named retromolar cavity. Pus can accumulate both between the periosteum and bone, and can spread over loose fiber of the front surface of the lower jaw in the direction of palatine tonsils causing their inflammation.

Abscesses of sublingual roller and jaw tongue groove. Sublingual roller, plica sublingualis, is formed by sublingual salivary gland enveloping loose fiber and coated with a thin mucusa.

Abscess causes compression of salivary gland located in sublingual space of lingual nerve, artery and vein. Due to the abscess development lingual artery becomes separated from all these elements and pressed to the sublingual-lingual muscle. The infection can spread into the cellular spaces of sublingual and submandibular regions, pterygo-mandibular cellular space.

Phlegmon is an acute diffuse purelent inflammation of cellular spaces, unlike abscess phlegmon has no precise boundaries.

Patotopography. ***Phlegmon of the temporal region*** can be a result of dissemination of inflammatory infiltrates from pterigo-mandibular or peripharyngeal spaces, infratemporal, retromandibular and pterigopalatine fossae and from the buccal region. Phlegmon can be located between the skin and temporal aponeurosis, surface and deep sheets of temporal fascia, and under the deep sheet of temporal fascia. In all this cases the phlegmon is squeezed by the temporal muscle. When phlegmon is located between the skin and temporal aponeurosis the compression takes place in the medial direction of a. temporalis superficialis, n. auriculotemporalis н n. intraorbitalis, which are located in the subcuteneous tissue. In interaponeurotic or subaponeurotic phlegmon the compression takes a place in a. temporalis profunda, n. temporalis profundi, a. meningea media, as the temporal muscle penetrates to the inner surface of the chewing muscle and borders with outside surface of the inner pterygoid muscle. It makes conditions for the pus to spread under the chewing muscle and towards perypharingeal space.

Phlegmon of perypharyngeal space often occurs as a result of the infection from the side of the palatine tonsils when they are inflamed, as well as purelent-inflammatory processes in the phlegmons located in the submandibular, sublingual and pterygo-mandibular spaces. As the perypharingeal space is connected with anterior mediastenum by neurovascular bundle the infection can spread to the mediastenum. The infection can spread along lingual artery to the bottom of the mouth.

Phlegmon of mouth bottom is a common purelent-inflammatory process exciting two or more cellular spaces located higher or lower than the diaphragm of the bottom of the mouth. The bottom of the mouth or its bottom wall is formed by soft tissues unit located between the tongue and hyoid bone in which inflammatory processes often develop. Here the lingual and upper tyroid arteries and veins, sublingual nerve and branches of facial nerve are located and they are squeezed when phlegmons form.

As the mouth bottom communicates with the cellular space of tongue root an infection can spread to the tongue root as well as to the perypharyngeal space and then anterior mediastinum.

Parotitis (Fig. 33) is a purelent inflammation of the parotid gland that is accompanied by an increase of gland. Lobular structure of the gland predisposes to the migratory nature of inflammation. In purulent parotitis the abscess often occurs through the external auditory canal, that is due to the lack of a fascia on the upper edge of the gland and the adherence of the gland to the external auditory canal where pus breaks out through incisura cartilaginis meatus acustici. If the internal plate of the fascial gland is broken the pus can penetrate into the parapharyngeal space and then along pharynx and esofagus to the posterior mediastenum with the development of mediastenitis.

Patotopography of parotitis: the mucous membrane of the oral cavity around the parotid duct is full, swollen, edema extends to the cellulose of the parotid and neck. Increased parotid glands can press the neurovascular bundle (that consists of external carotid artery and vein, n. vagus, glossopharyngeal, additional nerves), to the pre-vertebral muscles. The facial nerve passes through the thickness of the gland and its compression leads to the paralysis of mimical muscles.

Pannus (Fig. 34) is a vascularized granulated tissue penetrating into tissues under the epithelium. Pannus is formed due to the work of synovial cells that form a granulation tissue. Immunocompetent cells that produce an angiogenic factor

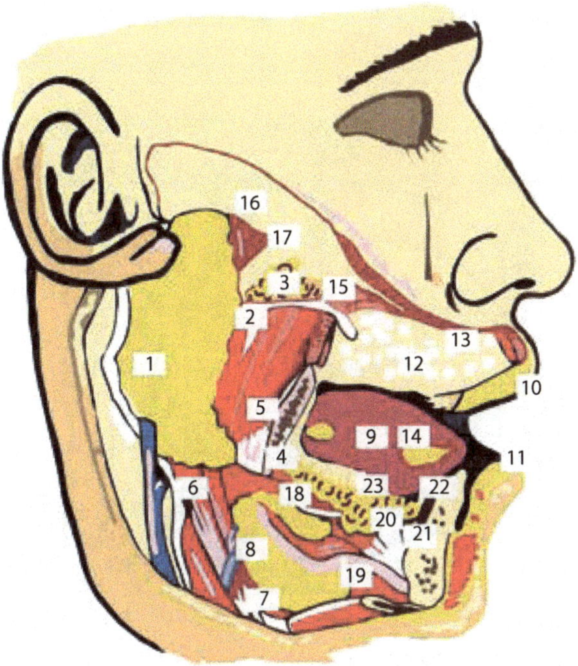

Figure 33 Purulent parotitis

1 –glandula parotis; 2 –ductus parotideus; 3 –glandula parotis accessoria; 4 –mandibula; 5 –m.masseter; 6 –m.digastricus; 7 –m. stylohyoideus; 8 –glandula submandibularis; 9 –lingua; 10 –labium superius; 11 –labium inferius; 12 –glandula buccalis; 13 –glandula labialis; 14 –glandula lingualis anterior; 15 –m.bussinator; 16 –fascia parotidea; 17 –fascia masseterica; 18 –ductus submandibularis; 19 – m.mylohyoideus; 20 –glandula sublingualis; 21 –m.genioglossus; 22 – ductus sublingualis major; 23 – ductus sublingualis minor

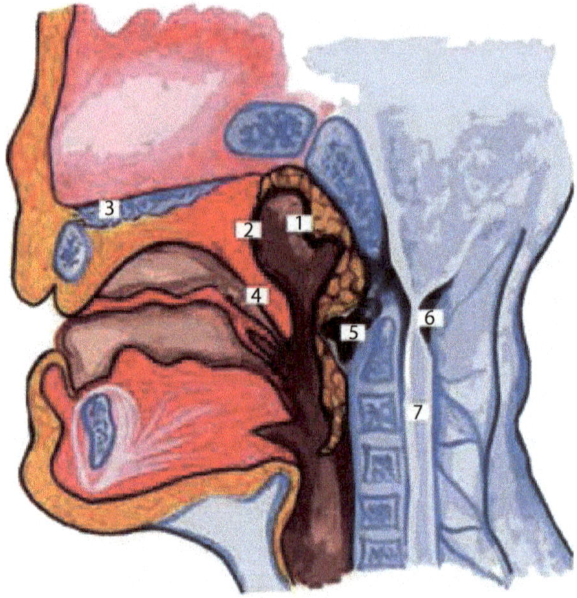

Figure 34 Pannus formation

1 - paries superior; 2 - ostium pharyngetim tubae auditivae; 3 - palatinum durum: 4 - palatinum molle; 5 – pannus; 6 – depression of spinal cord by odontoid process; 7 - medulla spinalis.

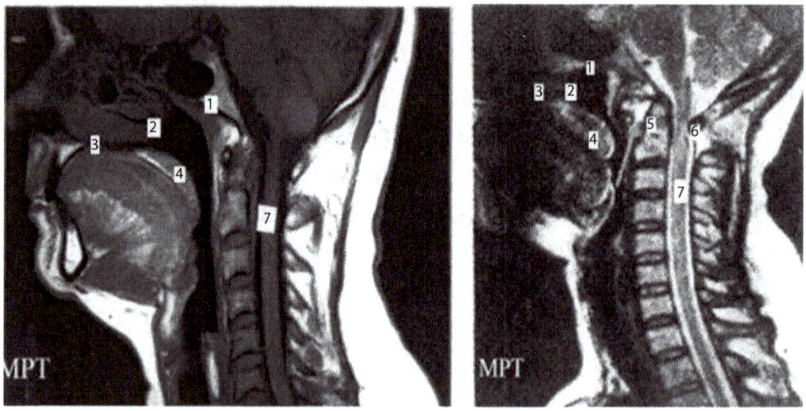

Figure 35 Pannus of nasopharynx

1 – paries superior; 2 – ostium pharyngeum tubae auditivae; 3 – palatum durum; 4 – palatum molle; pannus formation; 6 - depression of spinal cord by odontoid process; 7 – medulla spinalis.

cause the stimulation of synovial cell proliferation. This factor contributes to the growth of blood vessels in the cartilages. A magnetic resonance tomogram shows the formation of pannus - infiltration with neoplasm of connective tissue and development of vessels in this region. Because of the pannus, the spinal cord is compressed by the tooth process of the sinus through the dental fossa. In the average size of the maxillary sinus, its bottom is approximately at the level of the nasal cavity bottom, but often it is lower.

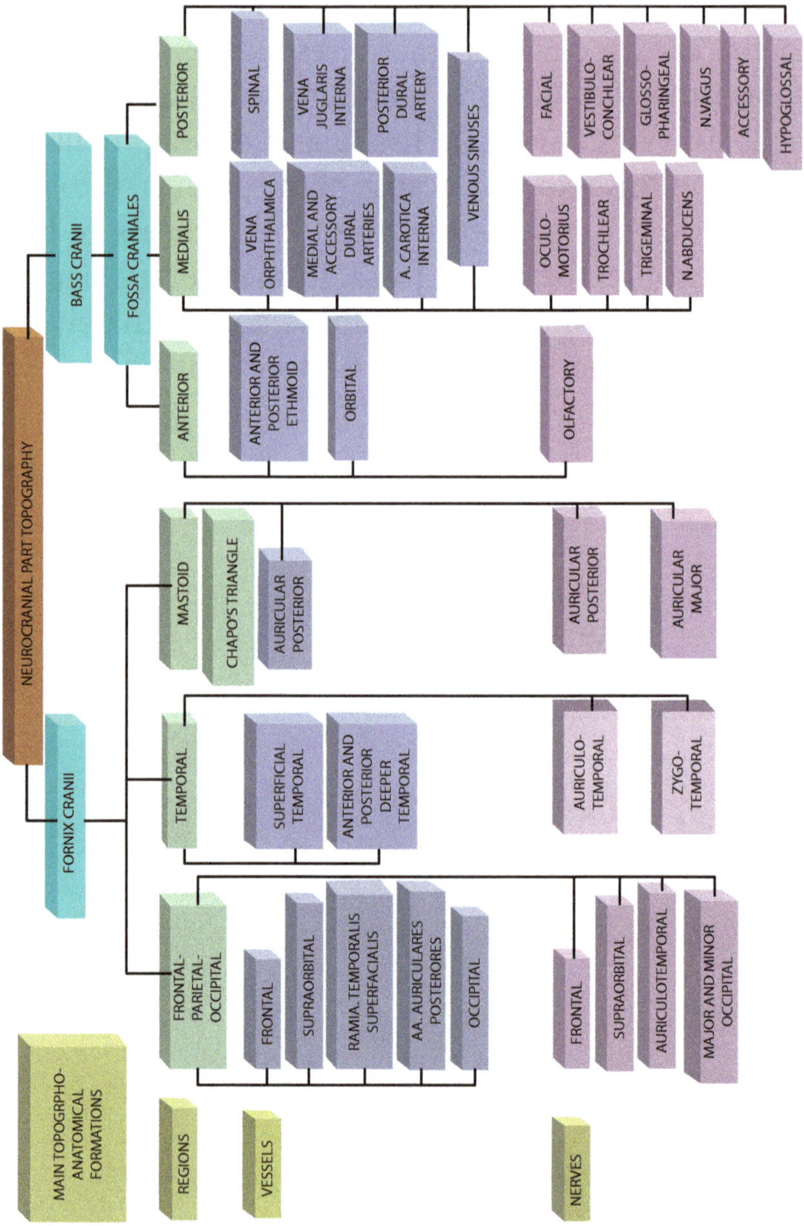

NEUROCRANIAL PART TOPOGRAPHY
BASS CRANII
FOSSA CRANIALES
POSTERIOR
SPINAL
VENA JUGLARIS INTERNA
POSTERIOR DURAL ARTERY
VENOUS SINUSES
FACIAL
VESTIBULO-CONCHLEAR
GLOSSO-PHARINGEAL
N.VAGUS
ACCESSORY
HYPOGLOSSAL
MEDIALIS
VENA ORPHTHALMICA
MEDIAL AND ACCESSORY DURAL ARTERIES
A. CAROTICA INTERNA
OCULO-MOTORIUS
TROCHLEAR
TRIGEMINAL
N.ABDUCENS
ANTERIOR
ANTERIOR AND POSTERIOR ETHMOID
ORBITAL
OLFACTORY
FORNIX CRANII
MASTOID
CHAPO'S TRIANGLE
AURICULAR POSTERIOR
AURICULAR POSTERIOR
AURICULAR MAJOR
TEMPORAL
SUPERFICIAL TEMPORAL
ANTERIOR AND POSTERIOR DEEPER TEMPORAL
AURICULO-TEMPORAL
ZYGO-TEMPORAL
FRONTAL-PARIETAL-OCCIPITAL
FRONTAL
SUPRAORBITAL
RAMIA. TEMPORALIS SUPERFACIALIS
AA. AURICULARES POSTERORES
OCCIPITAL
FRONTAL
SUPRAORBITAL
AURICULOTEMPORAL
MAJOR AND MINOR OCCIPITAL
MAIN TOPOGRPHO-ANATOMICAL FORMATIONS
REGIONS
VESSELS
NERVES

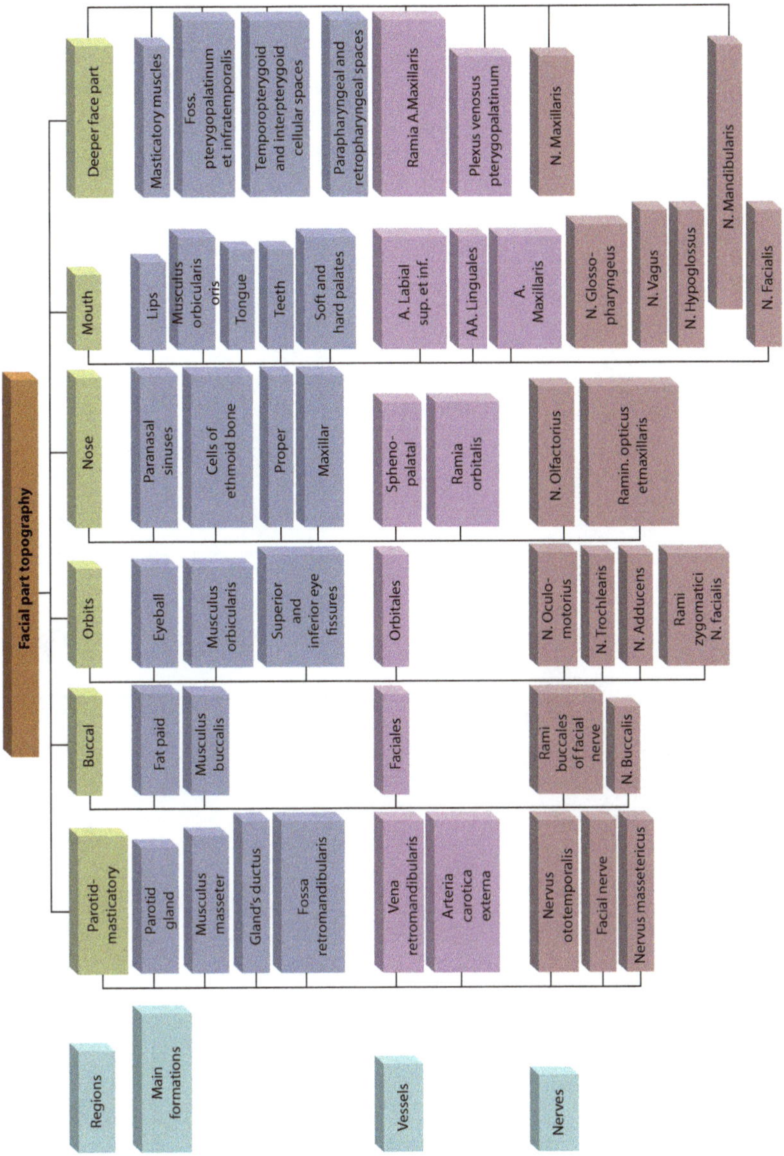
Facial part topography
Regions
Main formations
Vessels
Nerves
Parotid-masticatory
Parotid gland
Musculus masseter
Gland's ductus
Fossa retromandibularis
Vena retromandibularis
Arteria carotica externa
Nervus ototemporalis
Facial nerve
Nervus massetericus
Buccal
Fat paid
Musculus buccalis
Faciales
Rami buccales of facial nerve
N. Buccalis
Orbits
Eyeball
Musculus orbicularis
Superior and inferior eye fissures
Orbitales
N. Oculo-motorius
N. Trochlearis
N. Adducens
Rami zygomatici N. facialis
Nose
Paranasal sinuses
Cells of ethmoid bone
Proper
Maxillar
Spheno-palatal
Ramia orbitalis
N. Olfactorius
Ramin. opticus etmaxillaris
Mouth
Lips
Musculus orbicularis oris
Tongue
Teeth
Soft and hard palates
A. Labial sup. et inf.
AA. Linguales
A. Maxillaris
N. Glosso-pharyngeus
N. Vagus
N. Hypoglossus
N. Mandibularis
N. Facialis
Deeper face part
Masticatory muscles
Foss. pterygopalatinum et infratemporalis
Temporopterygoid and interpterygoid cellular spaces
Parapharyngeal and retropharyngeal spaces
Ramia A.Maxillaris
Plexus venosus pterygopalatinum
N. Maxillaris

Part 2: The Neck

Topographic Anatomy of the Neck

Topographic of neck *(Attachment 3)*

The boundary between the head and neck is the hyoid bone, os hyoideum. Above it - suprahyoid region, regio suprahyoidea, which belongs to the head and below - subhyoid region, regio infrahyoidea, which refers to the neck. The boundary line runs from the border to the corners of the bones of the lower jaw, goes around the external auditory canal and mastoid and goes back up to the upper nuchal line and protuberantia occipitalis externa is found with a similar line on the opposite side.

The lower limit of the neck is on the handle of the sternum manubrium sterni, clavicle, scapula acromion process and then coming to the spinous process of VII cervical vertebra.

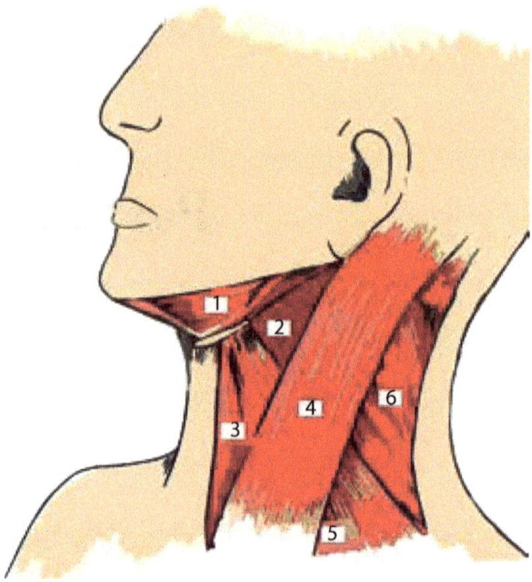

Figure 36 Neck area

1 – trigonum submandibulare; 2 – tr. caroticum; 3 – tr. omotracheale;4 – regio sternocleidomastoidea; 5 – tr. omoclaviculare; 6 – tr. omotrapezoideum.

External reference points and divide by the area (Fig. 36). The neck is divided into a front region, regio cotti anterior, and posterior region, regio cervicis. Within the anterior neck surgery is performed most frequently, as this is where most of the major organs in the neck are located.

Individual and age differences. Specific features depend on the constitution of the individual, for example, hypersthenics neck tight, short, for the most part, even thick, on the underside of the characteristic shape of the jaw has a sharp demarcation have astenikov long neck circumference is relatively small, cylindrical shape. Torticollis, sometimes developing in early childhood, degeneration occurs when the fascial capsule and contraction of m. sternodeidomastoideus, which is associated with intrapartum injury of this muscle.

Skeletopy and the projection of organs and neurovascular bundles on the surface of the skin. Larynx (larynx) is located at the level of the lower edge of C4 to C6. Windpipe (trachea) is located on the lower edge of the C6 to the upper edge Th5, where the bifurcation of the trachea. Throat (pharynx) is located on the base of the skull to the bottom of C6. Esophagus (oesophagus) extends from the lower edge of C6, passes through the thoracic cavity and terminates in the abdominal cavity at the level of Th11. Thyroid (glandula thyreoidea) - lateral lobes are located at the level of the larynx, and the isthmus of the gland lies in front of the trachea at the level from the first to the third of its cartilage.

Parathyroid glands (glandulae parathyroidea) four in number, are arranged between the capsule and the fascial sheath thyroid on the rear surface of its side lobes. Upper glands lie at the lower edge of the cricoid cartilage, the lower one transverse finger above the lower pole of the lateral lobes of the thyroid gland.

Fasciae, Superficial and Deep Cellular Spaces and their Relationship with Spaces Adjacent Regions (Fig. 37)

According to V.N Shevkunenko on the neck fascias to distinguish between 5:

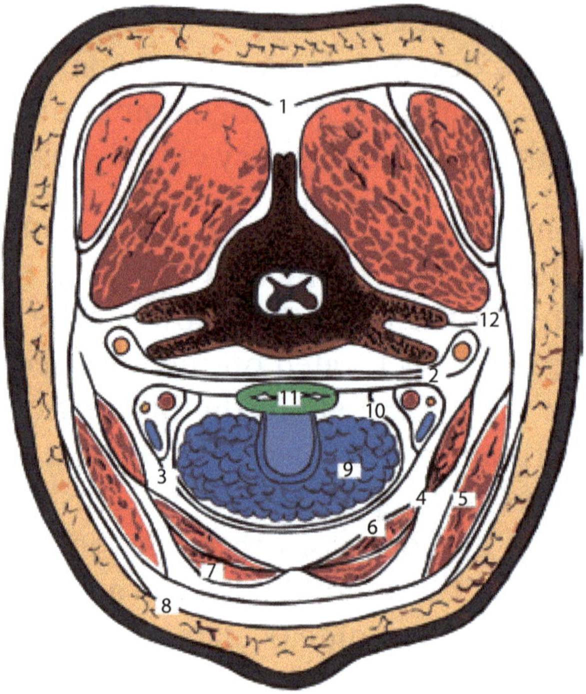

Figure 37 Fascias of neck

1 – second fascia; 2 – a second spur fascias; 3 – parietal leaf fourth fascias; 4 – third fascia; 5 - m. sternocleidomastoideus; 6 – m. sternothyroideus; 7 – m. sternohyoideus; 8 – first fascia; 9 – gl. thyroidea; 10 – visceral layer of the fourth fascias; 11 – esophagus; 12 - Spur of the fascia of the neck

- First fascia (fascia superficialis) - lies in its leaflets m. platysma. In violation of the innervation of the muscles of the neck becomes flabby appearance.
- The second fascia (fascia colli propria) - forms the fascial sheath for m. trapezius, m. sternodeidomastoideus and submandibular salivary glands. Slightly closed fascial vagina m. sternodeidomastoideus limits inflammation caused, for example,

in connection with apical mastoiditis and break-through of pus in the muscle.

- Third fascia (fascia colli media).

- Fourth fascia - visceral (fascia endocervicalis), covers all the organs of the neck, is divided into visceral and parietal leaflets and forms the neuro-vascular space, which houses a. carotis communis and n. vagus. Lower down the space communicates with the anterior mediastinum, which can move the inflammatory processes arising in the neck.

- Fifth fascia - prespinal (fascia prevertebralis), separates the anterior part of the neck from behind. The sheets of the fascia is truncus sympathycus three cervical nodes. In front of the fascia is the fifth cellular spaces, which extends to the level of pharyngeal lymphatic ring Pirogov-Valdeyra and down behind the esophagus and trachea, according to the posterior mediastinum.

Posterior to the neck of - between the fourth and vertebral (fifth) fascia of the neck - is behind the visceral cellular spaces, spatium retroviscerale. On either side of the neck organs are enclosed in a common fascial sheath common carotid artery, internal jugular Vienna, the vagus nerve and the deep lymph nodes of the neck. This so-called cellular spaces of the neuro-vascular bundle.

At the back of the throat abscess purulent process can spread along the loose fiber in the posterior mediastinum with development back mediastenitis so retropharyngeal abscesses are subject to urgent surgery. Behind the third fascia is pretracheal space communicating with the fiber behind the breastbone. Infection visceral fat before bed can lead to front mediastenitis. It is in this tissue can be injected air at the technical errors that arise when a tracheostomy is performed.

Reflex Zones

1. The main neurovascular bundle of the neck (common carotid artery, the vagus nerve and the internal jugular Vienna) - projection above;
2. Sinocoratid reflexogenic zone (bifurcation of the common carotid artery) - is projected on the upper edge of the thyroid cartilage 1 cm outwards;
3. Application of the sympathetic trunk: the top node is projected onto the transverse process of C3; Average unit is projected onto the transverse process of C6; cervicothoracic (stellate) node is projected at the level of the neck of the first rib;
4. cervical plexus;
5. brachial plexus;
6. The subclavian artery and brachial plexus trunks projected in the middle of the clavicle.

Lymphatic vessels and regional lymph nodes (Fig. 38). In the neck there are two groups of lymph nodes: front neck, nodi lymphatici cervicales anteriores, and lateral neck, nodi lymphatici servicales laterales.

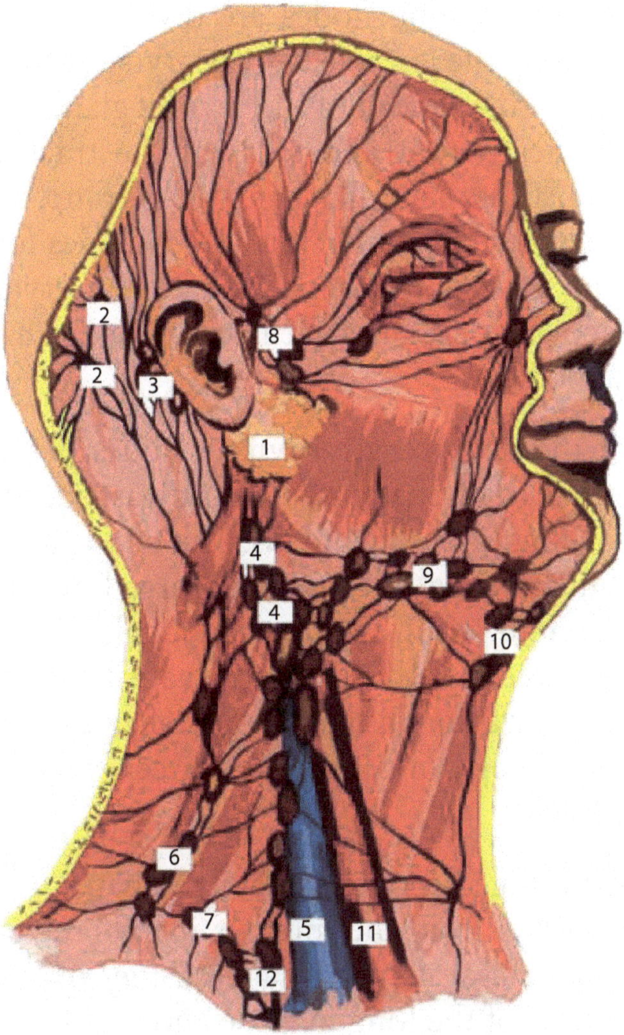

Figure 38 The lymphatic system of the head and neck

1 – gl. parotidea; 2 – l-di occipitales; 3 – l-di auriculares posteriors; 4 – l-di cervicales profundi superiors; 5 – v. jugularis profunda; 6 – l-di cervicales superficialees; 7 – l-di cervicales profundi superiors; 8 – l-di auriculares anteriores; 9 – l-di submaxillares; 10 – l-di submentales; 11 – a. carotica communa; 12 – truncus lymphaticus jugulares dextra

Anterior cervical lymph nodes are divided into superficial and deep, among the last release: prelaryngeal (lie in front of the larynx), thyroid (ahead of the thyroid gland), pretracheal and paratracheal (front and sides of the trachea). Lateral units also constitute superficial and deep group. Surface sites lie along the external jugular vein. Deep nodes form a chain along the internal jugular vein, the lateral artery of the neck (supraclavicular nodes) and the back of the pharynx - retropharyngeal nodes.

Because of the deep cervical lymph nodes deserve special attention nodus lymphaticus jugulo-digastricus and nodus lymphaticus jugulo-omohyoideus. The first is located on the internal jugular vein at the level of a large horn of the hyoid bone. The second lies in the internal jugular vein just above the m. omohuoideus. They take language lymphatic vessels, either directly or through the submental and submandibular lymph nodes. They can get the cancer cells when the tumor affects language.

The retropharyngeal nodes, nodi lymphatici retropharyngeal, lymph flows from the mucous membrane of the nasal cavity and paranasal its pneumatic cavities of the hard and soft palate, base of the tongue, nose and oropharynx, as well as middle ear. From all of these lymph nodes is flowing to the cervical nodes.

Lymph vessels:

- skin and muscles of the neck directed to nodi lumrhatisi servisales superficiales;

- larynx (lymphatic plexus mucosa above the vocal cords) - through the membrana thyrohyoidea to nodi lymphatici sevisales anteriores rrofundi; lymph vessels of the mucous membrane below the glottis are two ways: in front - through the membrana thurohuoidea to nodi lumrhatisi servisales anteriores rrofundi (predortannym) and posterior - to nodules located along n. laryngeus recurrens (paratracheal);

- thyroid - mainly to nodi lymphatici retropharyngeum et cervicales laterales (thyroid); the isthmus - the front surface of the cervical nodes;

- by the throat and tonsils lymph flows to nodi lymphatici retropharyngei et cervicales laterales profundi.

Triangles of the Neck

The inner (medial) triangle of the neck (trigonum cervicis mediale) (Fig. 39) limited medially median line, from the top - the bottom edge of the lower jaw, laterally -sternocleidomastoid muscle. The medial triangle isolated suprahyoid region (regio suprahyoidea) and subhyoid region (regio infrahyoidea).

Suprahyoid region (Fig. 40) has the shape of a triangle, two of its sides presented the base of the lower jaw, and the third - the hyoid bone and the posterior belly of the digastric muscle.

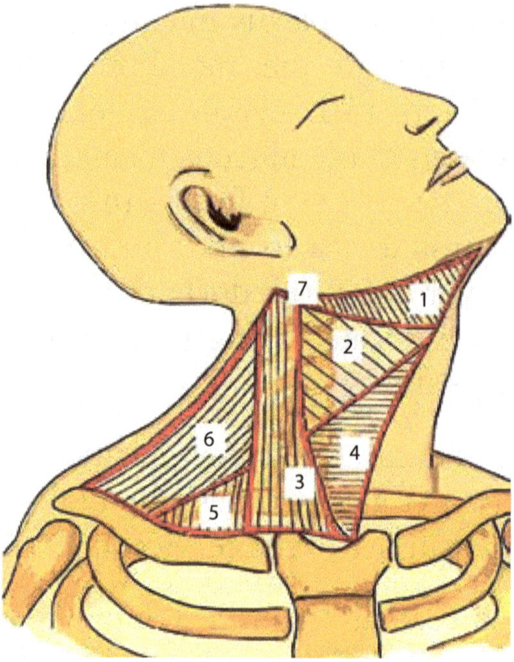

Figure 39 Areas of triangles and neck

1 – trigonum submaxillare; 2 – trigonum caroticum; 3 – regio m. sternocleidomastoidei; 4 – trigonum omotracheale; 5 – trigonum omoclaviculare; 6 – trigonum omotrapezoideum; 7 – fossa retromandibularis.

Layered suprahyoid topography of the area: The skin (cutis); Body fat (panniculus adiposus); The outer plate of the superficial fascia (lamina externa fasciae superficialis); Platysma (platysma); The inner plate of the superficial fascia (lamina interna fasciae superficialis); En neck fascia (fascia colli propria); Submandibular gland (glandula submandibularis); Digastric anterior belly (venter anterior m. Digastrici); Mylohyoid muscle (m. Mylohyoideus).

Within the area there are three suprahyoid triangle: submandibular triangle, lingual triangle, and submental triangle.

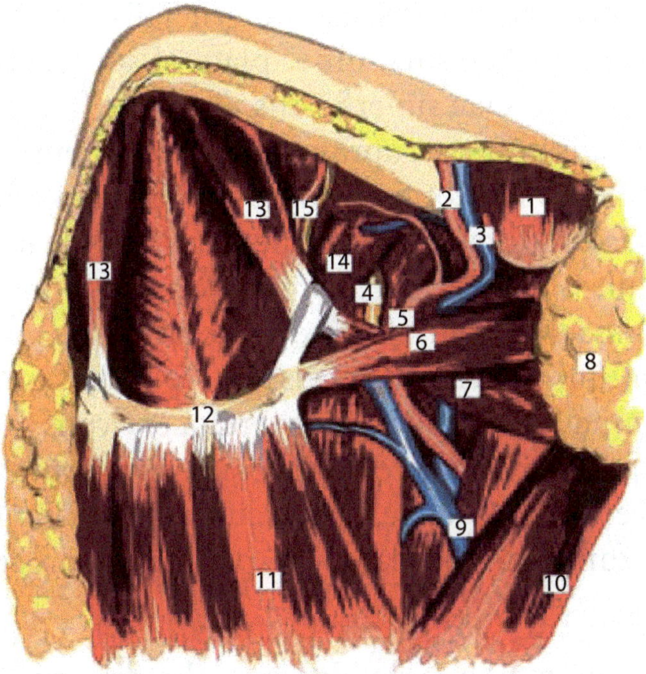

Figure 40 Deep layers suprahyoid area

1 – m. masseter; 2 – a. facialis; 3 – v. facialis: 4 – m. sublingualis; 5 – m. hyoglossus; 6 – m. stylohyoideus: 7 – venter posterior m. digastrici; 8 – gl. parotidea; 9 – v. facialis communa; 10 – m. sternocleidomastoideus; 11 – long muscles of the larynx; 12 – os hyoideum; 13 – venter anterior m. digastrici; 14 – m. mylohyoideus; 15 – a. mentalis et n. mylohyoideus

Subhyoid region (regio infrahyoid) occupies the lower part of the medial triangle of the neck. Borders subhyoid region: top - hyoid and posterior belly of digastric (venter posterior m. Digastiici), laterally and below - the front edge of the sterno-cleidomastoid muscle. Median line subhyoid area is divided into symmetrical halves.

Layered topography of the sublanguage area. Skin, cutis, fat deposits, panniculus adiposus, outer lamina of superficial fascia, lamina externa fasciae superficialis, subcutaneous neck muscle, m. Platysma, inner plate of superficial fascia, lamina interna fasciae superficialis, intrinsic fascia of neck, fascia cervicalis propria, suprahorn interperoneurotic space, spatium interaponeuroticum suprasternale, scapular-fascia fascia, fascia omoclavicularis, superficial muscle layer, stratum musculare superficiale, parietal lamina of intrasternal fascia, lamina Parietalis fasciae endocervicalis, vestibular space, spatium previscerale, visceral plate of intracereal fascia, lamina visceralis fasciae endocervicalis, posterior vascular space, spatium retroviscerale, invertebrate fascia, fascia prevertebralis, deep Th muscular layer, stratum muscularis profundum, cervical spine, pars cervicalis columnae vertebralis.

The sublingual region is divided into the drowsy and scapular-tracheal triangles, trigonum caroticum and trigonum omotracheale.

The outer (lateral) triangle (Fig. 41) is limited to the neck medially and top sternoclavicular-sosievidnoy muscle (m. Sternocleidomastoideus), below - the collarbone (clavicula), laterally - the trapezius muscle (m. Trapezius). The lateral triangle of the neck lower abdomen omohyoid muscle (venter inferior m. Omohyoidei) divided by scapuloclavicular and scapular-trapewievidny triangles (trigonum omoclaviculare et trigonum omotrapezoideum).

NVB inner and outer triangles of the neck (Fig. 41):

Common carotid artery (Fig. 41) (arteria carotis communis) - steam, originates in the chest cavity, right from the

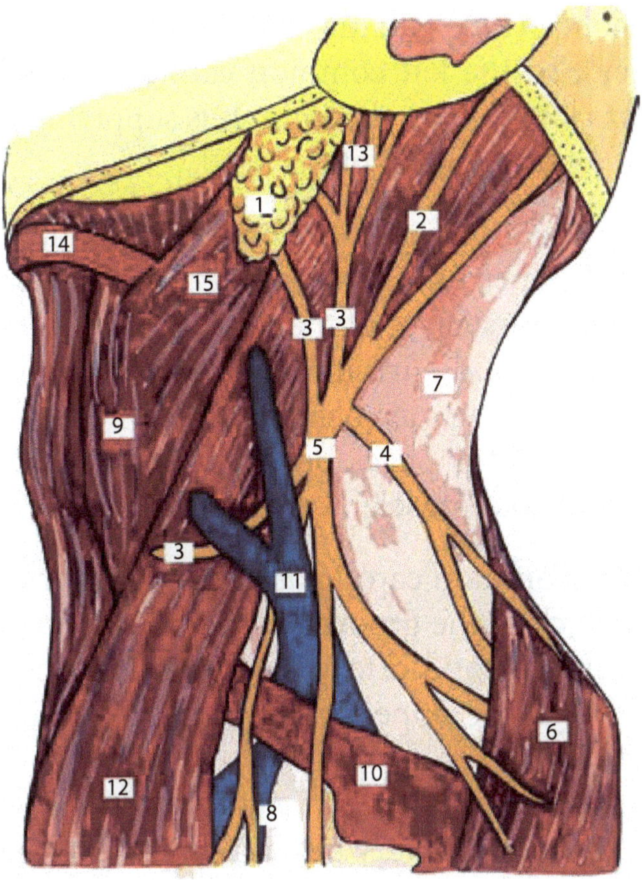

Figure 41 Side Triangle neck

1 – gl. salivaria parotidea; 2 – n. occipitalis minor; 3 – n. auricularis magnus; 4 – n. accessorius; 5 – nn. Supraclaviculares; 6 – m. trapezius; 7 – trigonum omotrapezoideum; 8 – trigonum omoclaviculare; 9 – venter anterior m. omohyoidei; 10 - venter posterior m. omohyoidei; 11 – v. jugularis externa; 12 – m. sternocleidomastoideus; 13 – n. cutaneus colli; 14 – venter anterior m. digastrici; 15 - venter posterior m. digastrici at stylohyoidei.

brachiocephalic trunk and the left - from the aortic arch (arcus aortae), so the left common carotid artery by a few centimeters is longer than the right. It is supplying the brain, the organ of sight and most of the head.

Common carotid artery rises almost vertically upwards and out through the apertura thoracis superior in the neck. Here it is on the anterior surface of the transverse processes of the cervical vertebrae and surrounding muscles on the side of the trachea and esophagus, behind the sternocleidomastoid muscle and fascia of the neck pretracheal plate with embedded in the omohyoid muscle.

Outside of the common carotid artery is the internal jugular Vienna, and back in the groove between the two - the vagus nerve. Common carotid artery in its course branches does not at the level of the upper edge of the thyroid cartilage is divided into:

- External carotid artery (arteria carotis externa),
- The internal carotid artery (arteria carotis interna).

External carotid artery (Fig. 42), arteria carotis externa, originates from the common carotid artery, at the level of the upper edge of the thyroid cartilage, directed upwards, passing behind the digastric (behind the rear abdomen) and stylohyoid. Then it falls in submandibular hole and enters into the body of the salivary gland. At the level of the neck of the articular process of the mandible external carotid artery is divided into the maxillary artery and superficial temporal artery.

External carotid artery is divided into 4 groups of branches (of topographic features).

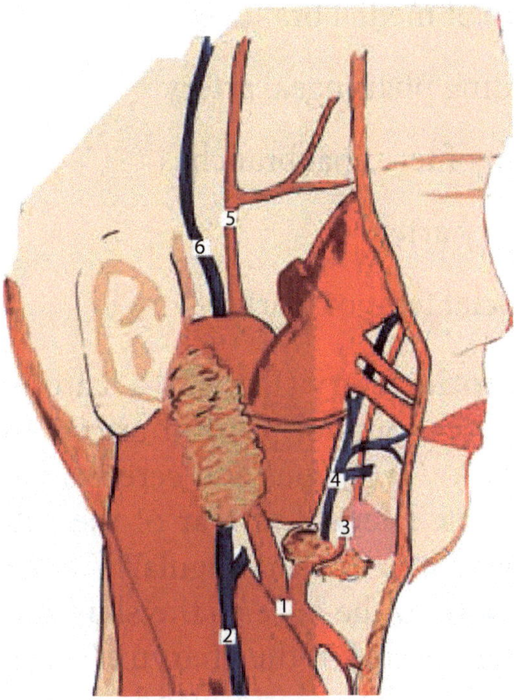

Figure 42 External carotid artery

1 – a. carotis externa; 2 – v. jugularis extrerna; 3 – a. maxillaris externa; 4 – v. facialis anterior; 5 – a. temporalis superficialis; 6 – v. temporalis superficialis

1. The group of front branches
   - Upper thyroid artery
   - Lingual artery
2. The group of rear branches
   - Sternocleidomastoid branch
   - Occipital artery
   - Posterior auricular artery

3. The group of medial branches
   - Ascending pharyngeal artery
4. The group of terminal branches
   - Maxillary artery
   - Superficial temporal artery

The internal carotid artery (Fig. 43), a. carotis interna, is a continuation of the common carotid artery. In it is distinguished neck, rocky, cavernous and cerebral part. Going up, it initially lies some lateral and rear of the external carotid artery. Laterally of it is internal jugular Vienna, v. jugularis interna. On its way to the base of the skull internal carotid artery runs along the side of the pharynx (cervical part, pars cervicalis) medial to the parotid gland, separated from the stylohypoglossal and stylopharyngeus muscle. In the cervical internal carotid artery branches usually do not give. Here it is somewhat expanded by the carotid sinus, sinus caroticus.

From the part of the brain a. carotis interna depart major artery.

I. ophthalmic artery, a. ophthalmica - doubles large vessel. On its way ophthalmic artery gives:

   1. lacrimal artery, a. lacrimalis,
   2. the central retinal artery, a. centralis retinae,
   3. short and long rear ciliary arteries, aa. ciliares posteriores breves et longae;

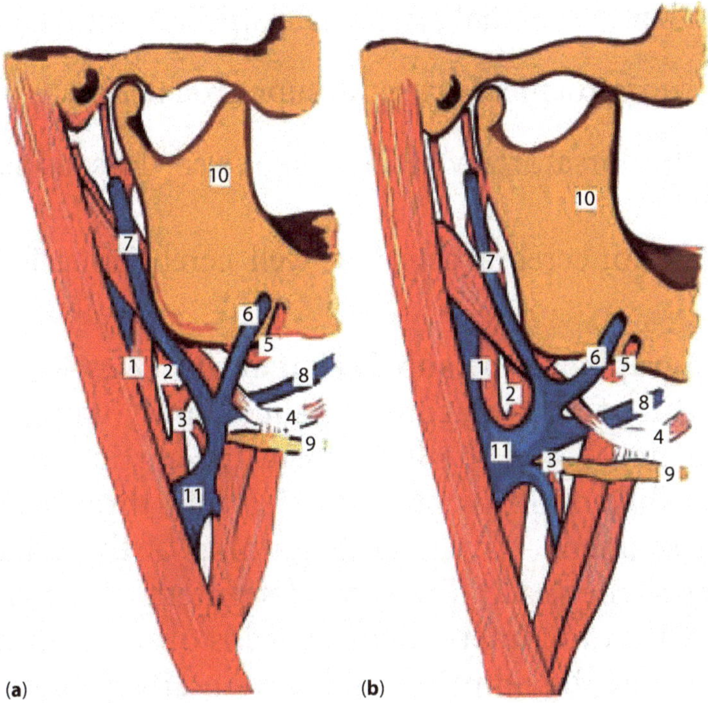

Figure 43 Options relations bifurcation of the common carotid arteries and veins

Bifurcation of the above veins Bifurcation covered vein

1 – a. carotica interna; 2 – a. carotica externa; 3 – v. thyroidea superior; 4 – m. digastricus; 5 - a. facialis; 6 - v. facialis; 7 – v.retromandibularis; 8 – v. lingvalis; 9 – os hyoideus; 10 – mandibula; 11 – v. jugularis interna.

4. muscular arteries, aa. musculares;
5. rear grid artery as well. ethmoidalis posterior;
6. anterior ethmoid artery, a. ethmoidalis anterior;
7. supraorbital artery as well. supraorbitals;

8. Century medial artery, aa. palpebrales mediales;
9. supratrochlear artery, a. supratrochlearis;
10. the dorsal artery of the nose as well. dorsalis nasi.

II. Anterior cerebral artery, as well. cerebri anterior.

The topography of the subclavian artery (Fig. 44). Leaves often right of the brachiocephalic trunk (truncus brachiocephalicus), and the left - from the aortic arch (arcus aortae). Conventionally, it is divided into three segments. The first segment - from the beginning of the arteries to the interscalene space. The second segment of the artery is located within the interscalene space is on I rib on it from the artery remains imprint - subclavian artery sulcus (sulcus a. Subclaviae). The third segment begins at the outlet of the interscalene space to the outer edge of the ribs I, which already begins axillary artery (a. Axillaris). The artery is arc-shaped. In the first otrez-ke it goes in superolateral direction lies horizontally in the second, and the third should be inclined downward.

Subclavian artery gives five branches: three in the first segment, and one in the second and third segments.

The branches of the first segment: vertebral artery (a. Vertebralis), Internal thoracic artery (a. Thoracica interna), thyrocervicalis trunk (truncus thyrocervicalis)

The branches of the second segment: From the second segment of the subclavian artery departs edge-cervical trunk (truncus costocervicalis) heading up the dome of the pleura

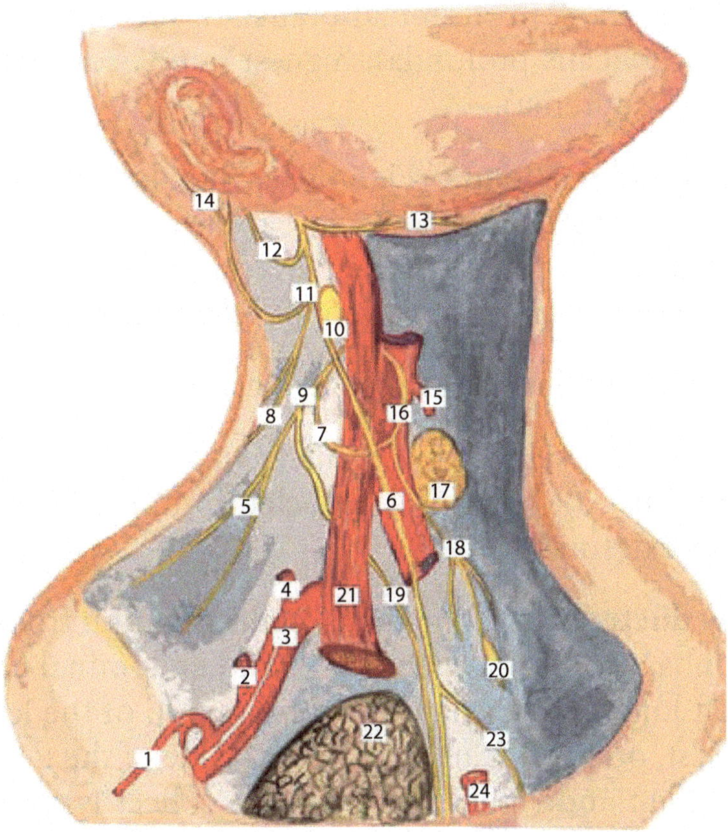

Figure 44 Innervation and blood supply of the neck muscles

1 – a. thoracoacromialis; 2 – a. transversa colli; 3 – a. subclavia; 4 – a. profunda colli; 5 – nn. supraclavicularis;6 – n. vagus; 7 – radix inferior ansae cervicalis;8 – n. transversus colli; 9 – IV n. spinalis cervicalis; 10 – ganglion sympathicus cervicalissuperior; 11 – III n. spinalis cervicalis; 12 – n. accessorius; 13 – n. hioideus; 14 - n. occipitalis minor; 15 – a. thyroidea superior; 16 – radix superior ansae cervicalis; 17 – gl. thyroidea; 18 – neck loop and its branches to the muscles located below the hyoid bone; 19 – n. phrenicus; 20 - ganglion sympathicus cervicalis medius; 21 – m. scalenus anterior;22 – apex pulmonis; 23 – n. laryngeus recurrens; 24 – a. thoracica interna

and dividing into two terminal branches: the deep cervical artery (a. cervicalis profunda); highest intercostal artery (a. intercostalis suprema).

External jugular Vienna (Fig. 45) (v. Jugularis externa) is formed at the angle of the lower jaw at the confluence of the posterior auricular vein (v. Auricularis posterior) and anastomosis with retromandibularis vein (v. Retromandibular).

The internal jugular Vienna (v. Jugularis interna)

Brachiocephalic veins: right and left, rr. hracliincephalicae, (lextru et sinistra), blood was collected from the head, neck and upper extremities.

Thoracic duct, ductus thoracicus, collects lymph from both lower limbs, organs and pelvic wall and abdominal cavities of the left lung, the left half of the heart wall of the left half of the chest, on the left arm and the left side of the neck and head. The left and right lumbar trunks collect lymph from the lower extremities, and the walls of the pelvis, abdomen, lumbar and sacral regions of the spinal canal and spinal membranes.

Phrenic nerve cervical plexus (n. Phrenicus) is formed from the anterior branches of the cervical spinal nerves. Initially, both the nerves are in the upper mediastinum, then move to the middle mediastinum and are located on the side of the pericardium, in front of the root of the corresponding lung. There phrenic nerve lies between the pericardium and mediastinal pleura and ends in the thickness of the diaphragm. Motor fibers innervate the diaphragm phrenic nerve, sensory

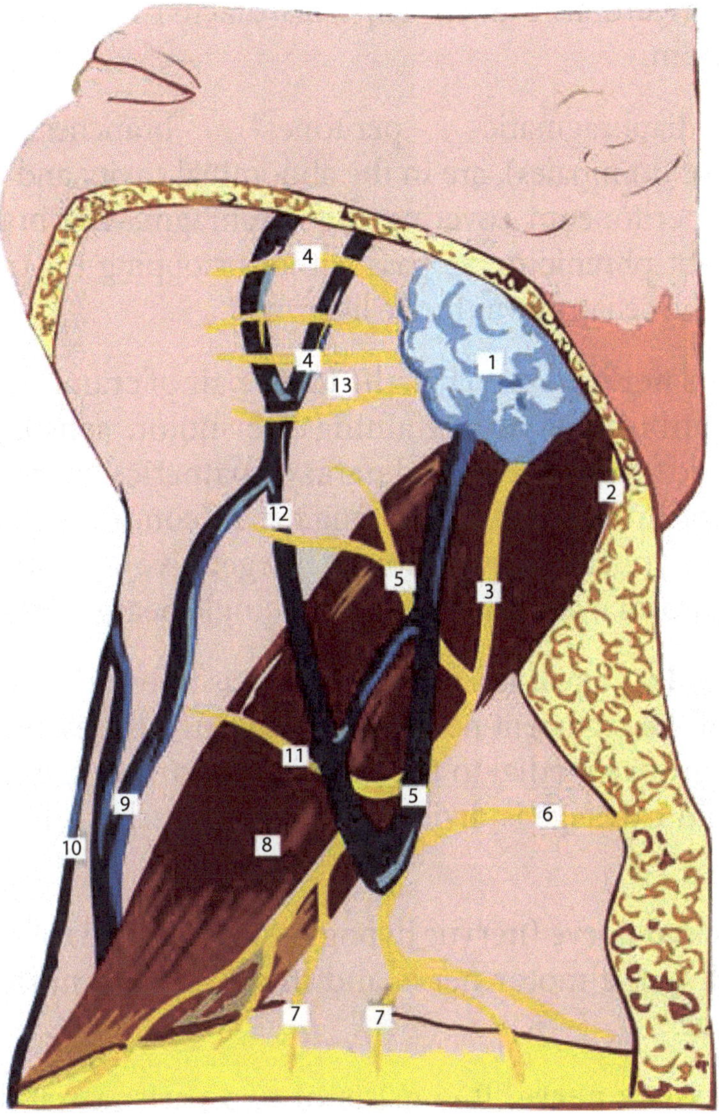

Figure 45 Surface vessels and nerves of the neck

1 – gl. salivaria parotidea; 2 – n. occipitalis minor; 3 – n. auricularis magnus; 4 – rr. n. facialis; 5 – v. jugularis externa; 6 – n. infraclavicularis posterior; 7 – nn. supraclaviculares anterior et medius; 8 – m. sternocleidomastoideus; 9 – v. jugularis anterior; 10 – v. mediana colli; 11 – n. cutaneus colli; 12 – anastomosis between the anterior and posterior jugular veins; 13 - gl. salivaria submandibularis

fibers pericardial branch (r.pericardiacus) - pleura and pericardium.

Sensitive-diaphragmatic peritoneal branches (rr. Phrenicoabdominales), are in the abdominal cavity and innervate the peritoneum covering the diaphragm. The branches of the right phrenic nerve pass without stopping (in transit), through the celiac plexus to the liver.

The vagus nerve in humans, the tenth pair of cranial nerves, doubles mixed nerve containing the motor, sensory and autonomic (sympathetic and parasympathetic) fibers. It has three core in the medulla oblongata, in common with the glossopharyngeal nerve: dorsal (vegetative), ventral, or double (motor), and the core t. N. a single beam (sensitive).

The lower laryngeal nerve (n. Laryngeus inferior) - the final branch of the recurrent nerve, passes through the esophageal-tracheal groove medial to the thyroid lobe and at the level of the cricoid cartilage is divided into two branches - the front and rear.

Hypoglossal nerve (nervus hypoglossus) - XII pair of cranial nerves. It is the motor nerve and innervates the muscles of the tongue.

The accessory nerve. It consists of two parts: a wandering and a lumbar. The conductive path is dvuhneyronnym motor.

Topography of the cervical plexus (Fig. 46-47). Cervical plexus (plexus cervicalis) formed by the anterior branches of the upper four cervical nerves. Upon emerging through the intervertebral foramen (foramen intervertebrale) these nerves lie on the front surface of the deep muscles of the neck

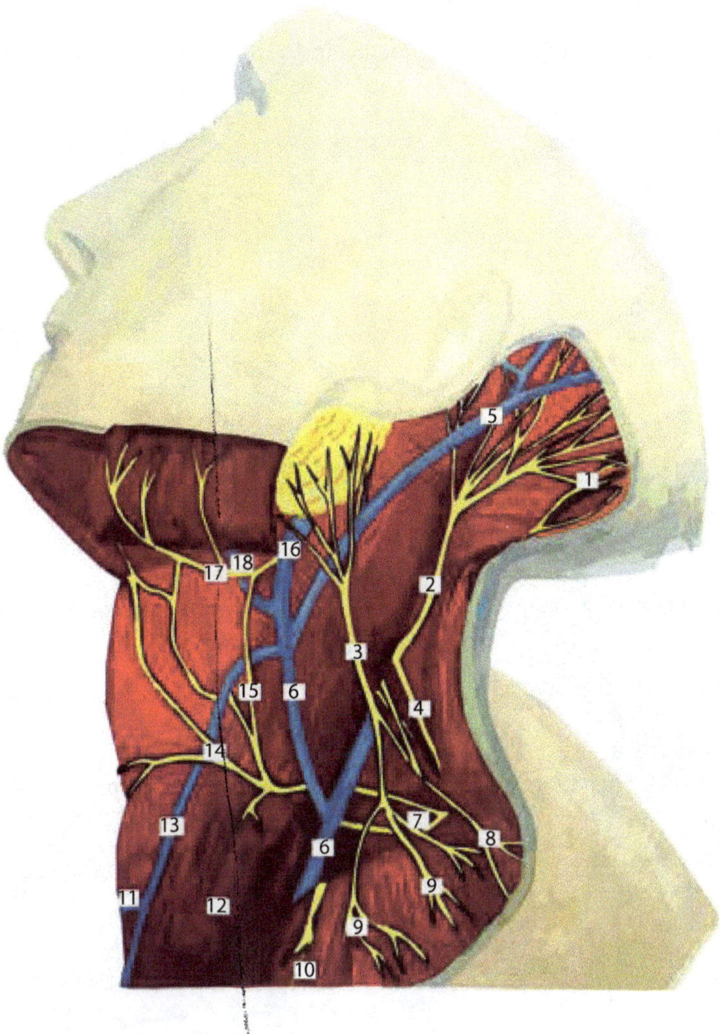

Figure 46 The branches of the cervical plexus

1 – v. occipitalis et n. occipitalis major; 2 – n. occipitalis minor; 3 – n. auricularis magnus; 4 – n. accessorius; 5 – v. auricularis posterior; 6 – v. jugularis externa; 7 – m. omohyoideus; 8 – nn. Supraclaviculares posteriores; 9 – nn. Supraclaviculares mediales; 10 - nn. Supraclaviculares anteriores; 11 – arcus venosus juguli; 12 – m. sternocleidomastoideus; 13 – v. jugularis anterior; 14 – n. cutaneous colli; 15 – facial nerve anastomosis with cutaneous nerve of the neck; 16 – v. retromandibularis; 17 – r. cervicalis n. facialis; 18 – v. facialis

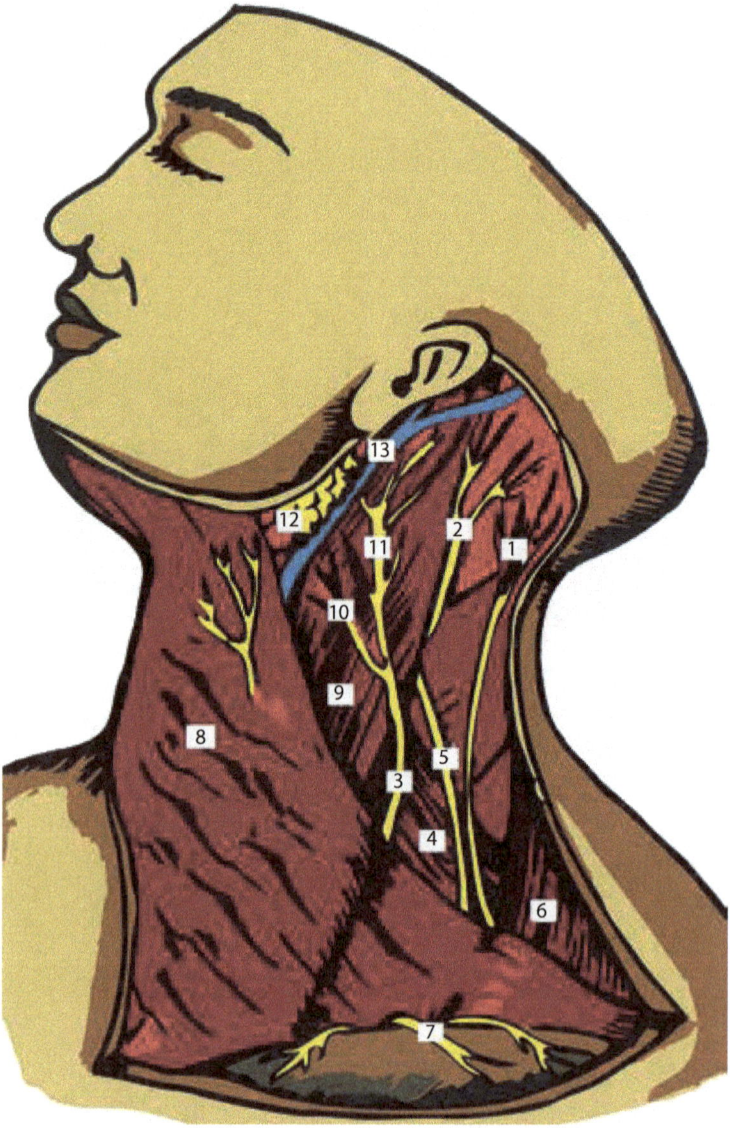

Figure 47 Surface education neck

2 – n. occipitalis minor; 3 – n. auricularis magnus; 4 – m. levator scapulae; 5 – n. accessorius; 6 – m. trapezius; 7 – nn. supraclaviculares; 8 – platysma; 9 – m. sternocleidomastoideus; 10 – r. anterior n. auricularis magnus; 11 – r. posterior n. auricularis magnus; 12 – gl. salivaria parotidea; 13 – v. retromandiularis

at the level of the upper four cervical vertebrae behind the sternocleidomastoid muscle.

Cervical plexus forms a sensory, motor (muscle) and mixed branches.

- Sensitive branches. Because of the sensitive nerves cutaneous branches formed neck (transverse nerve of neck, medial, intermediate and lateral supraclavicular nerves, great auricular nerve and small occipital nerve) described above.

- Motor branch. Motor branch of the cervical plexus (rami musculares plexus cervicalis) innervate the anterior, middle and posterior scalene muscles (mm. Scaleni anterior, medius et posterior), the long muscles of the head and neck (m. Longus capitis et colli), anterior and lateral rectus muscles of the head (m. rectus capitis anterior et m. rectus capitis lateralis), front intertransverse neck muscles (mm. intertransversarii anteriores cervicis) and levator scapulae muscle (m. levator scapulae). By the motor branches of the cervical plexus include cervical loop (ansa cervicalis). The neck loop is formed by connecting the upper spine (radix superior) and lower spine (radix inferior). Upper spine is formed from fibers I cervical spinal nerve joins the hypoglossal nerve, it extends in its composition to the external carotid artery. On the front surface of the external carotid artery to the bifurcation it descends and further on the front surface of the common carotid artery to connect to the lower spine. The lower back carries the fibers of the second and third cervical spinal nerves and leaves directly from the cervical plexus. After connecting the root of the neck loop branches to depart sternohyoid, sterno-thyroid, scapular-hyoid and thyrohyoid muscle (m. Stemohyoideus, m. Sternothyroideus, m. Thyrohyoideus et m. Omohyoideus).

- Mixed branch. Mixed nerve carrying both motor and sensory fibers, - paired phrenic nerve (n. Phrenicus). Phrenic nerve formed anterior branches C3-4, rests on the front surface of the anterior scalene muscle (m. Scalenus anterior), for it descends into the mediastinum and on the lateral surface of the pericardium (in front of the root of the lung) reaches the diaphragm. Right phrenic nerve through the opening of inferior vena cava (foramen venae cavae) enters the abdominal cavity and, after passing through the celiac plexus (plexus celiacus), participates in the innervation of the liver. From phrenic nerves depart: pericardial sensitive branch (r. Pericardiacus), innervating the pericardium and pleura; sensitive diaphragmatic-abdominal branches (rr. phrenicoabdominales), innervate the peritoneum lining the diaphragm; motor branches to the diaphragm.

The topography of the brachial plexus (Fig. 48-49). Brachial plexus (plexus brachialis) is formed from the anterior branches of the four lower cervical and thoracic spinal

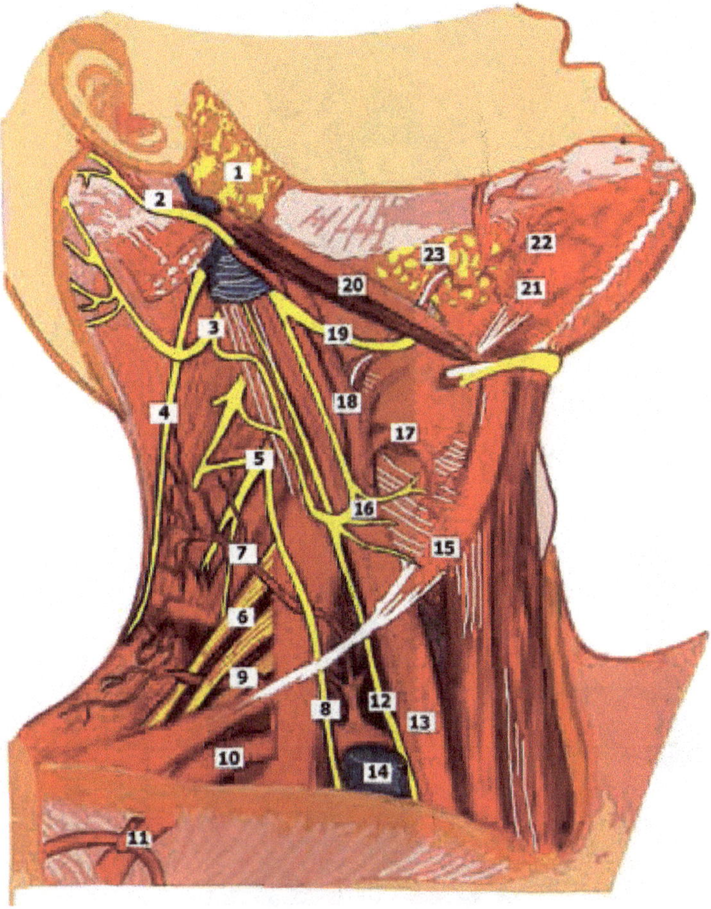

Figure 48 Deep neck Education

1 – gl. salivaria paratidea; 2 – a.et n. auricularis posterior; 3 - n. occipitalis minor; 4 - n. accessories; 5 – plexus cervicalis; 6 – plexus brachialis; 7 – a. superficialis colli; 8 – n. phrenicus; 9 – a. transversa colli; 10 – a. subclavia; 11 – blood vessels of the chest and shoulder process 12 – n. vagus; 13 – a. carotis communis; 14 – v. jugularis interna; 15 – m. omohyoideus; 16 – ansa cervicalis; 17 - a. thyroidea superior; 18 - a. carotis externa; 19 – n. sublingualis; 20 – m. stylohyoideus; 21 – venter anterior m. digastricus; 22 – a. mentalis; 23 – gl. submandibularis

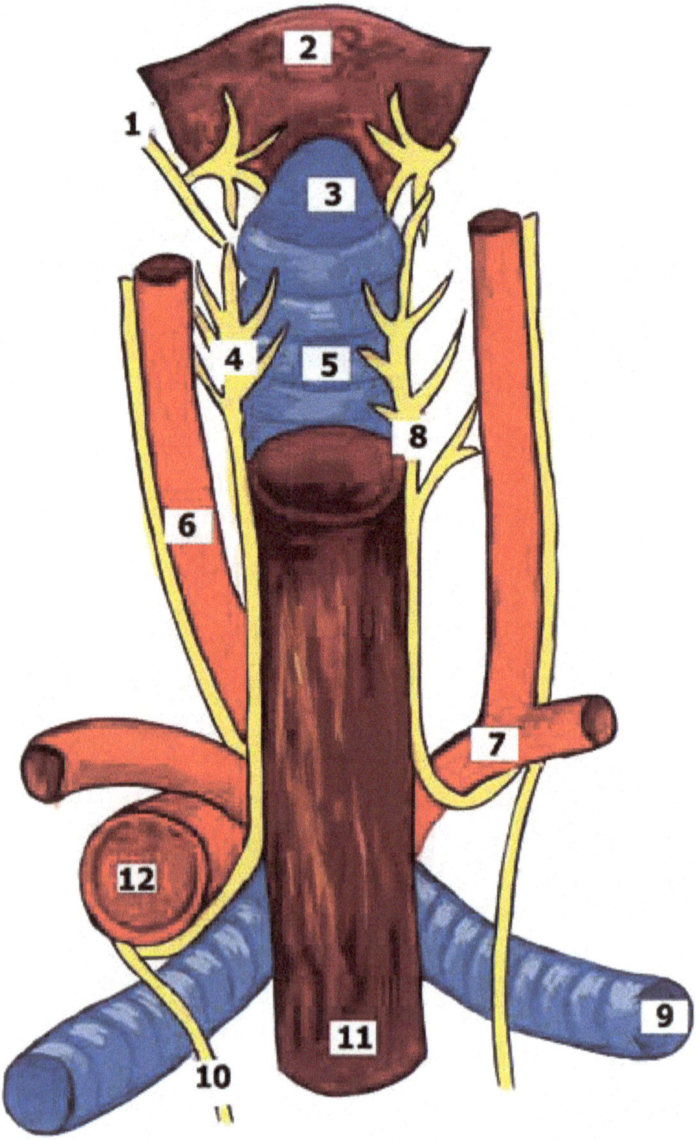

Figure 49 The topography of organs, blood vessels and nerves of the neck

1 – nn. laryngei superiores; 2 – basis lingvalis; 3 – epiglottis; 4 – nn. . laryngei inferiores; 5 – trachea; 6 – aa. caroticae communae; 7 – truncus brahiocephalicus; 8 – n. recurrens; 9 – bronchus; 10 – nn.vagi; 11 – esophagus; 12 – arcus aortae.

nerve 1 - C5-8, Th1. These five branches in the interscalene space form the trunks of the brachial plexus (trunci plexus brachialis).

- The upper trunk (truncus superior) is formed by fusion of the anterior branches of the fifth and sixth cervical nerves.
- Average trunk (truncus medius) - a direct continuation of the anterior branch of the seventh cervical nerve.
- Lower the trunk (truncus inferior) formed by the junction of the anterior branches C8 and Th1. The trunks of the brachial plexus interscalene space out in greater supraclavicular fossa (fossa supraclavicularis major), where the front of their separation (divisiones ventrales). The branches of a divided barrel again combine to form the lateral, medial and posterior bundles (fasciculus lateralis, fasciculus medialis et fasciculus posterior).

The brachial plexus is divided into two parts: the supraclavicular (pars supraclavicularis) and subclavian (pars infraclavicularis). Supraclavicular part of the brachial plexus at the exit of the interscalene space (spatium interscalenum) is located above the subclavian artery. Above the collarbone brachial plexus traverse laterally two arteries, above the surface passes cervical artery (a. Cervicalis superficialls), lower - vanes transversal artery (a. Transversa scapulae). Plexus passes between the barrel transverse cervical artery (a. Transversa colli).

The topography of the sympathetic trunk. Sympathetic trunk (truncus sympathicus) lies on the neck in front of the transverse processes of the cervical vertebrae from the neck to the base of the skull behind the ribs I or thicker prespinal fascias (fascia prevertebralis) on the front surface of the long muscles of the head and neck.

Sympathetic trunk in the neck often consists of the upper and middle cervical and cervicothoracic nodes (ganglion cervicale superior, ganglion cervicale medium et ganglion cervicothoracicum) and cross-site branches (rr. Interganglionares). The number of nodes varies from two to six.

Upper cervical node (ganglion cervicale superius) permanent mark. It has a spindle shape, a length of about 2 cm and a width of about 0.5 cm, located on level bodies II-III cervical vertebrae medial bottom unit of the vagus nerve.

In front of the upper cervical nodes are the internal carotid artery and internal jugular Vienna. The proximity of the superior cervical ganglion and the lower node of the vagus nerve level III cervical vertebrae allows you to vagosympathetic blockade on Vishnevsky.

Middle cervical node (ganglion cervicale medium) noted in 3/4 cases. It lies in the ladder-vertebral triangle (trigonum scalenovertebrale) at the level of the transverse process or Cv CVI higher arc inferior thyroid artery. From the middle of the cervical node depart following branches: Middle cervical cardiac nerve (n. Cardiacus cenncalis medius); connecting branches (rr. communicantes); cross-site branches (rr. interganglionares).

Cervicothoracic (stellate) node, ganglion cervicothoracicum (stellatum) show ever. It is formed at the confluence of the lower cervical to the first thoracic unit and located at the level of the transverse process of VII cervical vertebra behind the subclavian artery at the point of origin of the vertebral artery. Node flattened in the anteroposterior direction, a stellate form of its diameter of about 8 mm.

Organs of the Neck (Fig. 50–51)

Larynx (larynx) - part of the breathing apparatus is located between the hypopharynx and trachea at the level of the upper edge of the lower edge of the V to VI of the cervical vertebra. The larynx skeleton is formed of cartilage, ligaments and muscles. The larynx is associated with hyoid bone thyrohyoideus, membrane (membrana thyrohyoidea), stretched between the hyoid bone and the upper edge of the thyroid cartilage. Entrance to the larynx (aditus laryngis) is limited to the front of the epiglottis (epiglottis), laterally – arytenoid folds (plicae aryepiglottica), formed on the mucous membrane of the same name during the passage of the muscles and back - interaritenoide notch (incisura interarytenoidea). Over the entrance to the larynx (aditus laiyngis) is the cavity of the larynx (cavitas laryngis), which has three sections.

1. Vestibulum larynx (vestibulum laiyngis)
2. Middle part of the larynx - voice box itself (glottis)
3. Lower part of the larynx - cavitas infraglottica

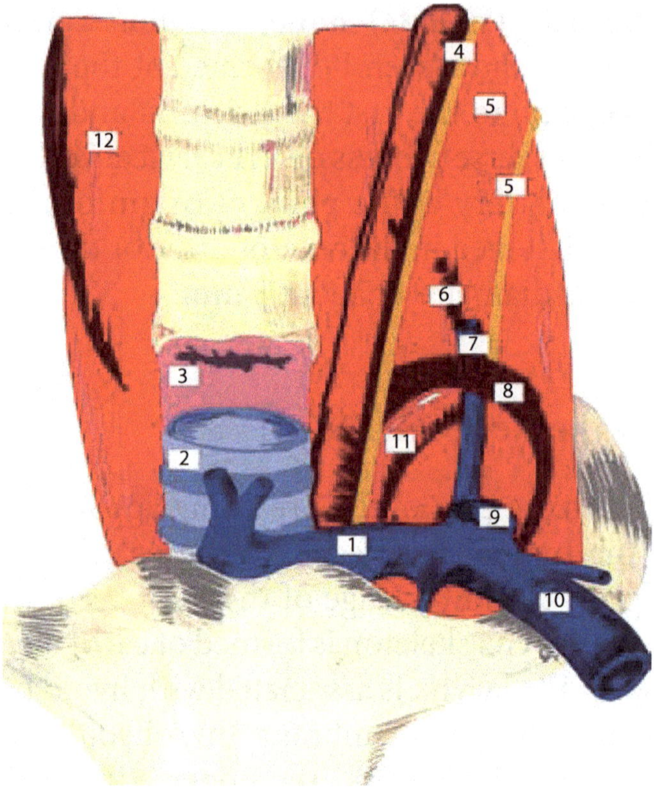

Figure 50 Topography staircase-vertebral triangle

1 – v. brachiocephalica sinistra; 2 – trachea; 3 – esophagus; 4 – a. caroticus communiset n.vagus; 5 – n. phrenicus et m. scalenus anterior; 6 – a. vertebralis; 7 – v. vertebralis; 8 – ductus lymphaticus thoracicus; 9 – v. jugularis interna; 10 – v. subclavia sinistra; 11 – a. subclavia sinistra; 12 – m. longus colli et capitis.

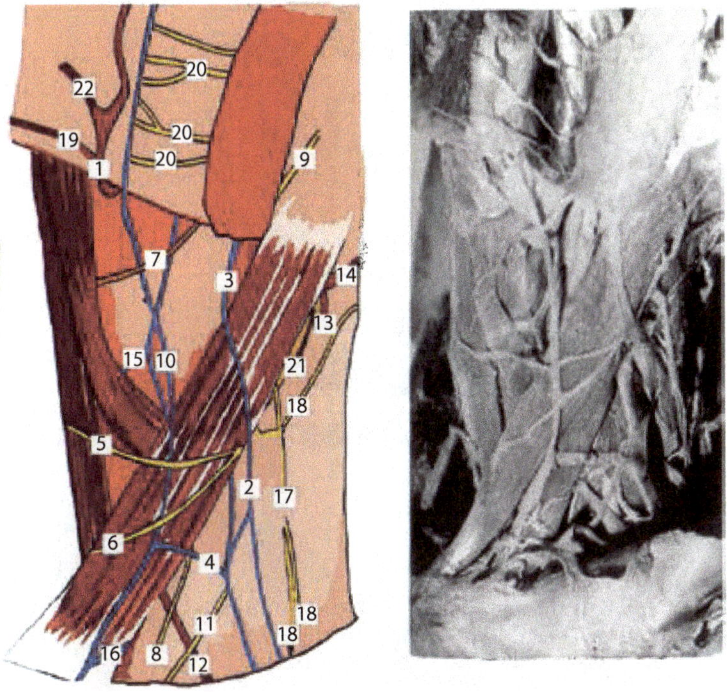

Figure 51

1 – a. facialis; 2 – vv. jugularis profunda; 3 – v.retromandibularis; 4 – anastomosis between the external jugular vein and the internal jugular vein; 5 – n. transversus colli superior; 6 – n. transversus colli inferior; 7 – r. cervicalis n.fascialis; 8 - n. accessories; 9 - n. occipitalis minor; 10 - v. jugularis externa; 11 – r. medialis n. supraclavicularis; 12 – a. transversa colli; 13 – a. occipitalis; 14 – a. auricularis posterior; 15 – v. transversa colli; 16 – v.musculi sternocleidomastoidei; 17 – r. lateralis n. supraclavicularis; 18 – rr. muscularis nervorum; 9 – a. submentalis; 20 – rr. bucales n. fascialis; 21 – n. auricularis magnum; 22 – r. labialis a.facialis.

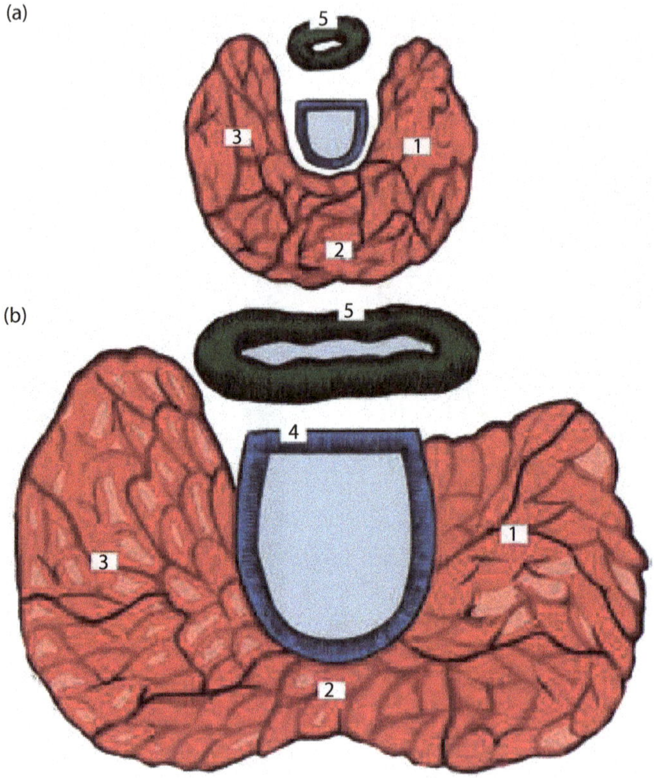

Figure 52 The cut at the level of the thyroid isthmus

Newborn: 1 – esophagus; 2 – trachea; 3 – lobus dexter et sinister; 4 – isthmus, 5 - esophagus Adult: 1 – esophagus; 2 – trachea; 3 – lobus dexter et sinister; 4 – isthmus, 5 - esophagus

Larynx consists of cricoid cartilage (cartilago cricoidea), thyroid cartilage (cartilago thyroidea), the two arytenoid cartilages (cartilago arytenoidea) and the epiglottis (epiglottis).

Blood supply of the larynx is carried out at the expense of the upper and lower laryngeal artery (a. Laiyngea superior et inferior). The first - a branch of superior thyroid artery, the second - the inferior thyroid artery. Venous outflow. Blood flow of the larynx occurs at the upper laryngeal vein (v. Laiyngea superior) to the top of the thyroid (v. Thyroidea superior) and further into the internal jugular vein (v. Jugularis interna), the lower laryngeal vein (v. Laryngea inferior) blood flowing to the bottom thyroid (v. thyroidea inferior) and later in brachiocephalic vein (v. brachiocephalica).

The innervation of the larynx occurs due to branches of the vagus nerve and sympathetic trunk of the upper node. From the upper assembly depart laryngopharyngeal 96 branches (rr. Laiyngo-pharyngei), penetrating the larynx as part of the upper laryngeal nerve and along the laryngeal arteries.

Lymph from the lower divisions of the larynx performed in front deep cervical lymph nodes (nodi lymphatici ceiyicales profundi anteriores) - prelaryngeal in the lymph nodes (nodi lymphatici prelaryngeales), thyroid lymph nodes (nodi lymphatici thyroidei), paratracheal lymph nodes (nodi lymphatici paratracheales). Lymphatic drainage of the upper sections of the larynx occurs in the lateral deep cervical lymph nodes (nodi lymphatici cervicales profundi laterals).

Trachea. It consists of 16-20 horseshoe-shaped cartilage (cartilagines tracheales), connected to each other annular ligament (ligamenta annularia). Behind the half-rings are connected to the mobile membranous tracheal wall (paries membranaceus). The trachea is divided into cervical (pars cervicalis) and thoracic (pars thoracica) parts.

Neck's part stretches from the beginning of the trachea at the lower edge of the VI cervical vertebrae to the thoracic inlet (apertura thoracis superior), which corresponds to the lower edge of ThII.

Neck's part of trachea is directed downward and backward at an acute angle. Therefore, the first of its cartilage lie no deeper than 1.5-2 cm from the skin at the level of the jugular notch (incisura jugularis) trachea lies at a depth of 4 cm. For this reason, the upper tracheostomy technically is easier than the lower tracheostomy.

Blood supply and venous drainage. Supplying the cervical part of the trachea branches of the inferior thyroid arteries (aa. Thyroideae inferiores), krovoottok occurs in the veins of the same name brachiocephalic vein (v. Brachiocephalica).

Innervation. Innervated by cervical part of the trachea branches of recurrent laryngeal nerve (nn. Laryngei reccurentes) and branches of the cervical sympathetic trunk.

The lymphatic drainage from the lower larynx is carried out in the front deep cervical lymph nodes, nodi lymphatici cervicales profundi anteriores, pre-laryngeal lymph nodes, nodi lymphatici prelaryngeales, thyroid lymph nodes, nodi lymphatici thyroidei, paratracheal lymph nodes, nodi lymphatici paratracheales. Lymph outflow from the upper parts of the larynx occurs in the lateral deep cervical lymph nodes, nodi lymphatici cervicalesprofundi lateralis

The esophagus (esophagus) is lined with mucous membrane muscular tube that connects the throat to the stomach. His clearance is given by the transverse slot size 1224 mm longitudinal folds mucosa. The esophagus is divided into a neck part (pars cenncalis), the breast area (pars thoracica) and the ventral portion (pars abdominalis).

The hypopharynx at the lower edge of the cricoid cartilage (which corresponds to CVI) moves up into the esophagus (esophagus). The cervical part of the esophagus (pars cervicalis esophagi) has a length of about 5 cm and passes into the breast area (pars thoracica) at the jugular notch of the sternum (incisura jugularis sterni), which corresponds to ThII. Daylight pharynx into the esophagus occurs in the median plane and has a constriction below the cervical esophagus lumen increases and there is a deviation from the esophagus to the left middle plane.

Blood supply and venous drainage. Supplying the cervical part of the esophagus branches of the inferior thyroid arteries (aa. Thyroideae inferiores), blood flow occurs in the veins of the same name brachiocephalic vein (v. Brachiocephalica).

Innervation. Innervating the cervical part of the esophagus branch of the recurrent laryngeal nerve (nn. Laryngei recurrentes) and branches of the cervical sympathetic trunk.

Lymph flow. The lymphatic vessels of the esophagus empty into the deep cervical lymph nodes (nodi lymphatici cervicales profundi).

The thyroid gland (Fig. 50). Thyroid (glandula thyroidea) relates to endocrine glands (glandulae endocrinae) produces hormones thyroxine, triiodothyronine (participating in the regulation of basal metabolism) and calcitonin (involved in regulation of the metabolism of calcium and phosphorus). The gland is located within the scapular-tracheal triangle (trigonum omotracheale), it consists of the right and left lobes (lobus dexter et lobus sinister), and the isthmus (isthmus glandulae thyroideae). Thyroid front cover sternohyoid, sterno-thyroid and omohyoid muscle (mm. Sternohyoideus, sternothyroideus, omohyoideus).

The isthmus of the thyroid gland in the front covers the trachea at the level of the top two of its cartilages, connecting the left and right lobe of the thyroid. The isthmus of the thyroid gland may be missing. In these cases, the thyroid gland can be conventionally considered paired organs. In a third case, a fraction of the pyramid (lobus pyramidalis), a conical ridge rising to the side plate of the thyroid cartilag.

Parathyroid gland. Parathyroid (parathyroid) gland (glandulae parathyroidea) - endocrine glands that produce parathyroid hormone involved in the regulation of calcium and phosphate. Parathyroid glands are elongated or rounded education brownish-pink length of 4-8 mm and a width of 3-4 mm, ie the size of a small pea. Number of parathyroid glands varies from 1 to 8. They are usually two pairs - top and bottom.

The upper parathyroid glands (glandulae parathyroideae superiores) lie between the fibrous capsule of the thyroid gland and visceral fascia plate endocervicalis at the cricoid cartilage in the middle of the distance between the upper pole and the isthmus of the thyroid gland, adjacent to the rear surface of its shares.

The lower parathyroid glands (glandulae parathyroideae inferiores) are located at the lower pole on the rear surface of the thyroid lobe between the fibrous capsule and visceral fascia plate endocervicalis in the area where entering the lower thyroid artery. In order to preserve the parathyroid glands by removing the thyroid gland - should separate out the bottom part of the thyroid gland, while retaining all the "panicle" vessels— ramifications of the inferior thyroid artery (a. thyroidea inferior).

Submandibular gland. Is a complex alveolar-tubular gland secreting and eyecatching mixed secret. It is covered with a thin capsule and is located in the submandibular triangle. Outside to the iron surface plate are adjacent cervical fascia and the skin. The medial surface of the gland adjacent to the sublingual, lingual and styloglossus. The upper edge of the iron in contact with the inner surface of the body of the mandible and the lower part is getting out of its lower edge.

By the submandibular gland suitable arterial branches of the facial artery and venous branches diverging in the same vein. Lymphatic vessels are based on the adjacent submandibular lymph nodes.

Afferent innervation of the prostate made fibers lingual nerve (of the mandibular nerve - the third branch of the trigeminal nerve, V pair of cranial nerves).

Afferent innervation of the gland is carried out by the fibers of the lingual nerve (From the mandibular nerve - the third branch of the trigeminal nerve, V pair of cranial nerves).

Efferent innervation is provided by parasympathetic and sympathetic fibers. Parasympathetic postganglionic fibers pass through the facial nerve (VII pair of cranial nerves) through a tympanic string and a submandibular jaw. Sympathetic fibers pass to the gland of the plexus around the external carotid artery.

Submandibular salivary gland. Superficial fascia in the submandibular triangle forms a holder for the platysma. The tissue between the platysma and 2nd cervical fascia neck branch of the facial nerve and the upper branch of n. transversus colli of the cervical plexus form the arcus cervicalis superficialis, located at the level of the hyoid bone. Fascia surrounds iron freely without fusing with it and not giving in depth cancer processes. Between the iron and its fascial bed there is a layer of loose fiber. With this submandibular gland can be easily separated from the bed by a blunt. The upper part of the outer surface of the gland adjoins directly to the periosteum of the mandible; internal (deep) rests on the surface of the iron mm. mylohyoideus et hyoglossus, separated from them by a deep lear second fascia.

Pathography of the Neck

The abscess (Fig. 53) is the formation limited by an infiltrative capsule, inside of which is a cavity containing purulent exudate. Abscess does not tend to spread to surrounding tissues

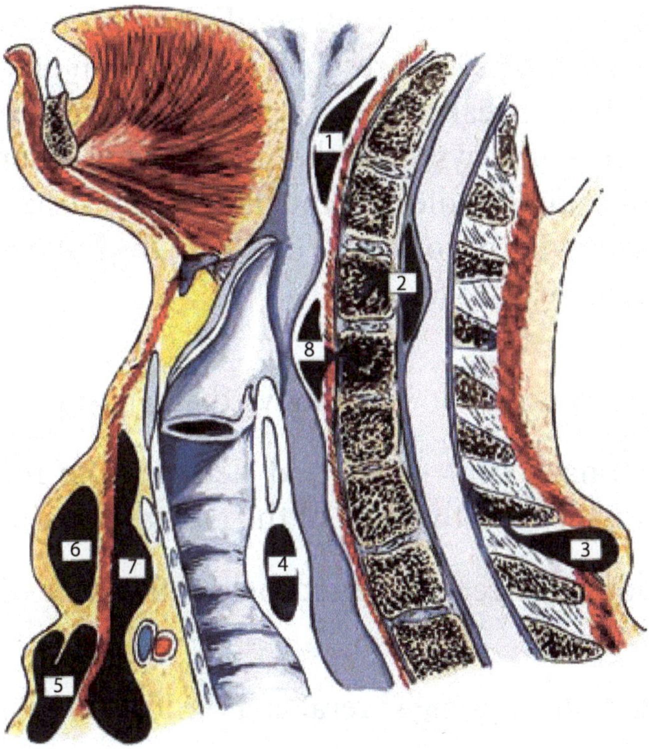

Figure 53 Typical localization of abscesses and abscesses of the neck

a - sagittal section: 1 - retropharyngeal abscess; 2 - extradural abscess; 3 - abscess nuchal region; 4 - retrotrahealny presternal abscess abscess; 6 - interapneuroticum episternal abscess; 7 - abscesses previsceral space; 8 – abscess behind esophagus

Pathotopography (Fig. 54). **Hyperal abscess** (retropharyngeal abscess) is formed as a result of suppuration of the lymph nodes and pharyngeal pharynx space. Pathogens penetrate the lymphatic pathways from the nasal cavity, nasopharynx, auditory tube and middle ear. The pharyngeal abscess is classified according to the location. There are the following types of retropharyngeal abscesses:

- Epipharyngeal - located above the palate
- Mesopharyngeal - localized between the root of the tongue and the edge of the palatal curtain
- Hypopharyngeal - located below the root of the tongue
- Mixed - occupying several anatomical zones.

The abscess is located in the pharyngeal space, which is located behind the pharynx. Limited to the rear of the prevertebral fascia, in front of theparapharyngeal fascia, from the sides to the pharyngeal-vertebral fascial spurs. At the top it starts from the base of the skull, below it passes into the cellulose located behind the esophagus (the back cerebral cell space of the neck), the latter passes into the fiber of the posterior mediastinum. In addition to cellulose, the pharyngeal space contains single lymph nodes.

With the formation of retropharyngeal abscesses, the purulent process can quickly spread along the course of loose fiber, into the posterior mediastinum with the development of a dangerous posterior mediastinitis.

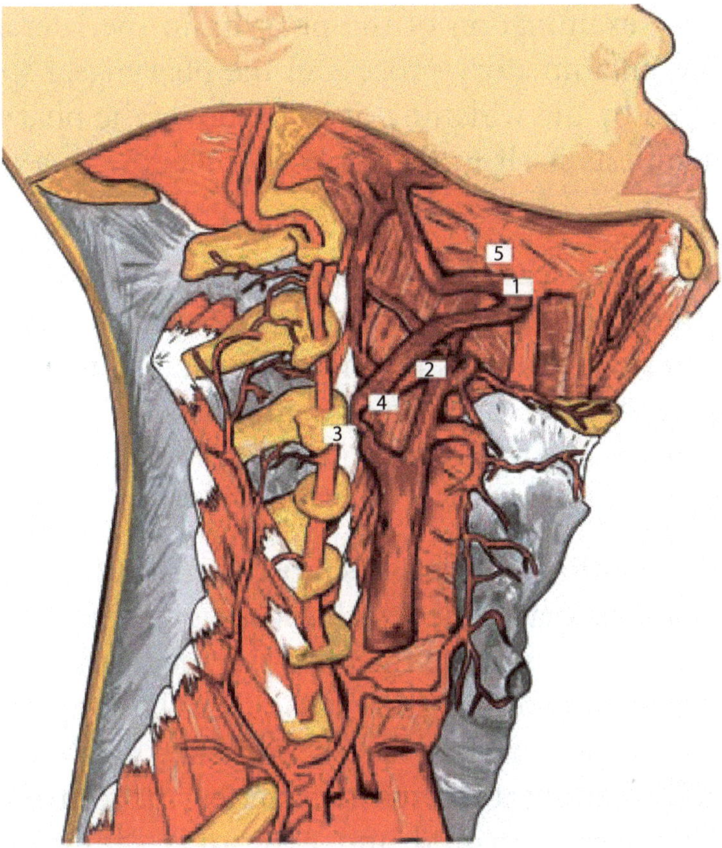

Figure 54 Patotopographic anatomy of the neck

1 – a. carotica interna; 2 – a. carotica externa; 3 – stenosis a. carotica internae; 4 – Crossing a. carotica interna et a. carotica externa; 5 – m. digastrici venter posterior.

In the X-ray examination of the pharynx in the lateral projection, the inflammatory process in the pharyngeal space is characterized by the widening of its shadow; The pharyngeal abscess manifests itself in the form of a shadow in a certain area.

Retro-tracheal abscess is an inflammatory process, localized between the trachea and esophagus. The abscess blends the trachea anteriorly, as a result, the narrowing of its lumen is observed. Also, the esophagus is compressed. Purulent exudate compresses the recurrent guttural nerves, the lower thyroid arteries.

Behind the esophagus abscess is most often located in a slotted retrovisceral space filled with loose fiber and spreading from the base of the skull to the posterior mediastinum up to the diaphragm. There is a compression of the long muscles of the neck, the prevertebral fascia. The abscess also squeezes the thoracic duct, right intercostal arteries, the terminal sections of the semi-unpaired and additional semi-unpaired veins.

Epidural abscess - an accumulation of pus between the dura mater and the bone. The epidural abscess develops more often in the middle chest and lower lumbar regions, where the epidural space is best expressed. There is a shift of the dura mater of the spinal cord forward. The formation of an abscess leads to compression of the spinal roots, and then of the spinal cord.

Pre-tracheal abscess is an accumulation of pus between the parietal and visceral sheets of the 4th fascia. With a massive inflammatory process, it is possible to squeeze the main neurovascular bundle of the neck surrounded by vagina carotica, which is formed by the parietal leaf of the 4th fascia. An abscess causes the thyroid gland to move forward. It is also possible to involve it in the inflammatory process.

Figure 54 depicts the pathological tortuosity of the right internal carotid artery, the right external carotid artery, which arises as a result of stenosis of the right internal carotid artery. These vessels cross each other, as a result of which the right internal carotid artery shifts to the anterior region of the neck and partially covers the posterior abdomen of the digastric muscle; The right external carotid artery is displaced to the lateral processes of the cervical vertebrae, which is not observed in the norm, a. Carotis interna, departs from the common carotid artery at the level of the upper edge of the thyroid cartilage. It does not enter the cranial cavity through the sleep canal, but goes to the facial part of the skull, where it splits into terminal branches. In cases of absence of the internal carotid artery, the lack of blood supply to the brain is compensated by a much greater development of the corresponding arteries of the opposite hemisphere, as well as by the unusual development of vertebral arteries.

Ectopic location of various species of goiter (Fig. 55). Ectopia of the tissue of the thyroid gland is a condition in which tissues of the thyroid are located not only in the natural location of the gland, but also go beyond it.

Language ectopic manifestations are the most common type of anomalies in 90% of cases of this condition. Sublinguals

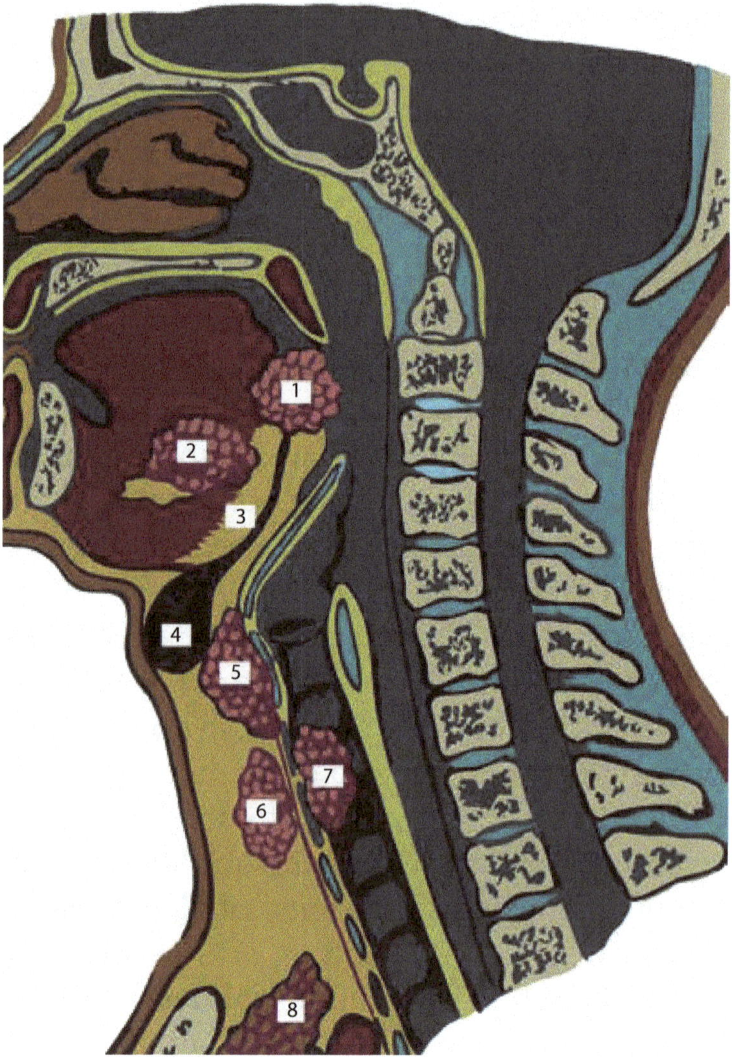

Figure 55 Ectopic and normal arrangement of the various types of goiter

1 - craw of the tongue; 2 - internal goiter; 3 - schitoyazychny goiter; 4 - cyst shchito-lingual duct; 5 - predgortanny goiter; 6 - normally located thyroid gland; 7 - intratracheal goiter; 8 - retrosternal goiter

are much less common. Such formations can be divided into the sublingual and sublingual or appear at the level of the hyoid bone.

Pathotopography. The tooth of the root of the tongue can develop both from dystopic and aberrant thyroid tissue. The disease occurs mainly in women. The root of the tongue is located mainly along the midline of the tongue in the region of the foramen caecum and rarely in the region of one of the halves of the tongue. The knot has a rounded shape, a smooth surface, a wide base, clear boundaries, located partly in the thickness of the tongue, unshifted. As a result of compression of the neighboring anatomical-topographical structures and formations, the goiter node can cause a change in the voice-the appearance, so-called, of the nasal voice, sensation of the foreign body in the pharynx, followed by the development of dysphagia or aphagia.

The cyst of the shield-lingual duct is located on the neck along the median line in the pre-tracheal space of the neck, it can be either front from behind the hyoid bone, in most cases the cysts are connected to the hyoid bone.

Retrosternal goiter is an abnormally low thyroid gland (below the jugular notch). There is compression and displacement of large bronchi, as well as large vessels. First of all, the veins are squeezed. Less commonly, the arteries are squeezed, namely, the subclavian and internal mammary arteries. There is compression of the nervous tables, the recurrent nerve, sympathetic trunk, brachial plexus and diaphragmatic nerve can be squeezed.

Intratracheal goiter is a goiter that forms in the uninfected thyroid duct, covered with the hyoid bone and thus located inside the tracheal wall. Goiter leads to a narrowing of the lumen of the trachea, thus causing respiratory failure. The node is usually located in the cervical region of the trachea at the back of the posterolateral wall, often on the left.

Nodal goiter - a group of thyroid gland diseases that occur with the development in it of volumetric nodal formations of various origin and morphology

Figure 56 shows the nodular goiter of the thyroid gland. When the goiter increases, it can squeeze the recurrent laryngeal nerve that lies outside the fascial capsule of the thyroid gland, as well as the main neurovascular bundle of the neck, which is located along the external sections of the gland, and the common carotid artery that is very closely connected with the thyroid gland, So it touches the fascial vagina of the arteries.

Behind and medially are the larynx, trachea esophagus, which can be squeezed by the nodular goiter of the thyroid gland (Fig. 56-57). When the larynx and trachea are compressed, the conduction of the upper respiratory tract is broken, which in turn leads to respiratory failure.

On ultrasound, the nodular goiter looks like a regular shape, isoechoic, there are cystic cavities, there are no hyperechoic inclusions, and there are also no vascular structures.

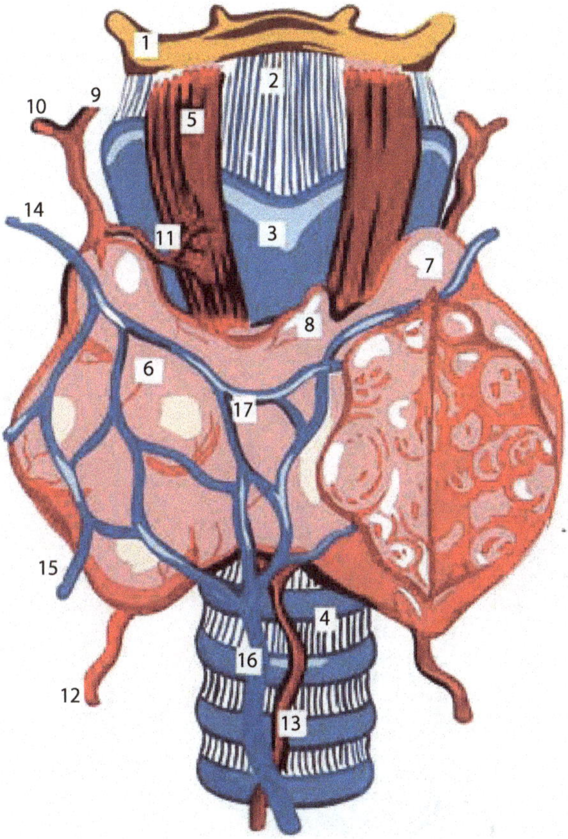

Figure 56 Nodular goiter of the thyroid gland

1 – os hyoideum; 2 – lig. thyrohyoideum; 3 – cartilago thyroidea;4 – trachea; 5 – m. thyrohyoideus; 6 – lobus sinister; 8 – lobus pyramidalis; 9 – a. laryngea superior; 10 – a. thyroidea superior; 11 – ramus cricothyroideus;12 – a. thyroidea inferior; 13 - a. thyroidea ima; 14 – v. thyroidea superior; 15 – v. thyroidea inferior; 16 – v. thyroidea media; 17 – plexus thyroideus impar

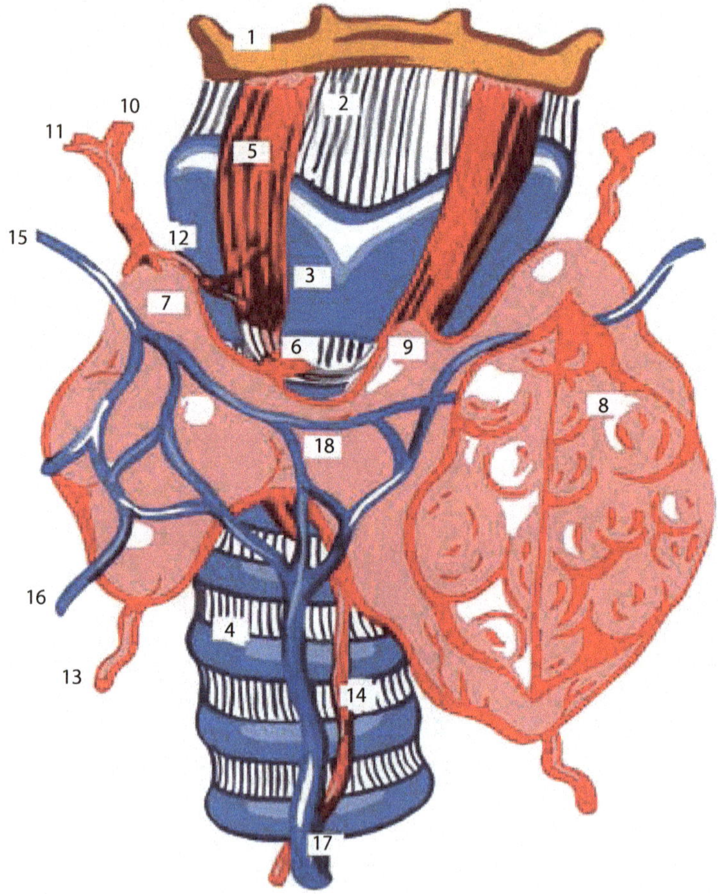

Figure 57 Thyroid nodular goiter with compression of the larynx and trachea

1 – os hyoideum; 2 – lig. thyrohyoideum medianum; 3 – cartilago thyroidea;4 – trachea; 5 – m. thyrohyoideus; 6 – m. cricothyroideus; 7 – lobus dexter; 8 – lobus sinister; 9 – lobus pyramidalis; 10 – a. laryngea superior; 11 – a.thyroidea superior; 12 – ramus cricothyroideus; 13 – a. thyroidea inferior; 14 – a. thyroidea ima; 15 – v. thyroidea superior; 16 – v. thyroidea inferior; 17 – v. thyroidea media; 18 – plexus thyroideus impar

Topography of the neck

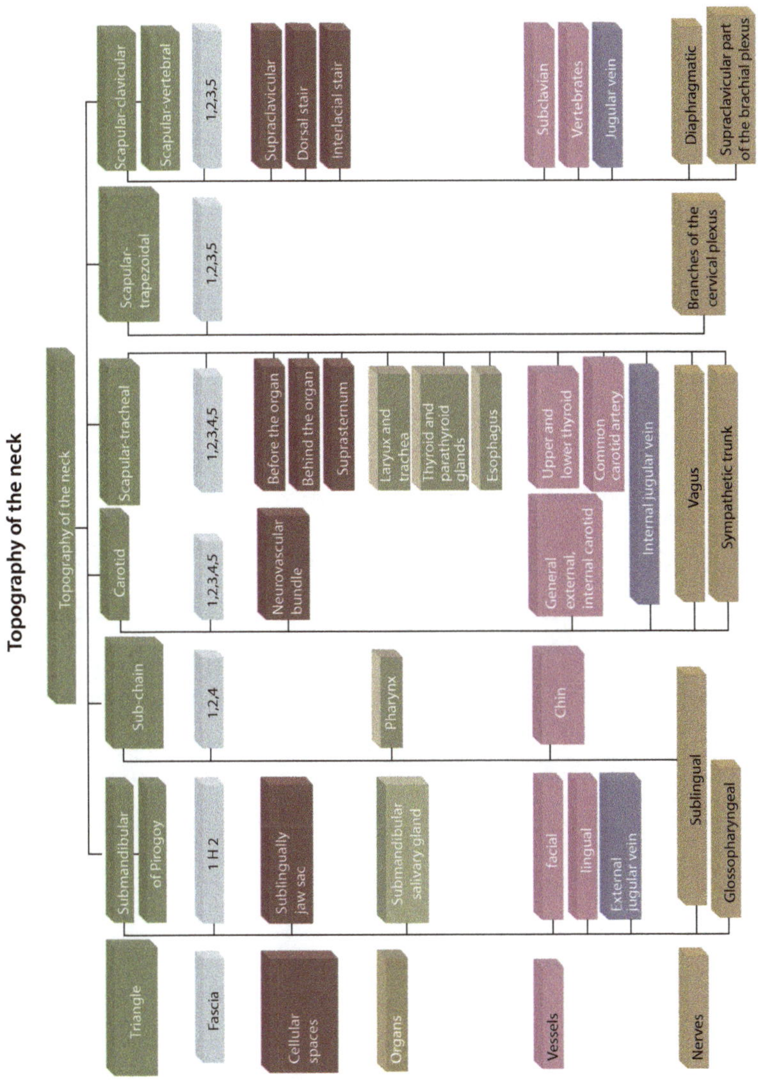

Appendix A

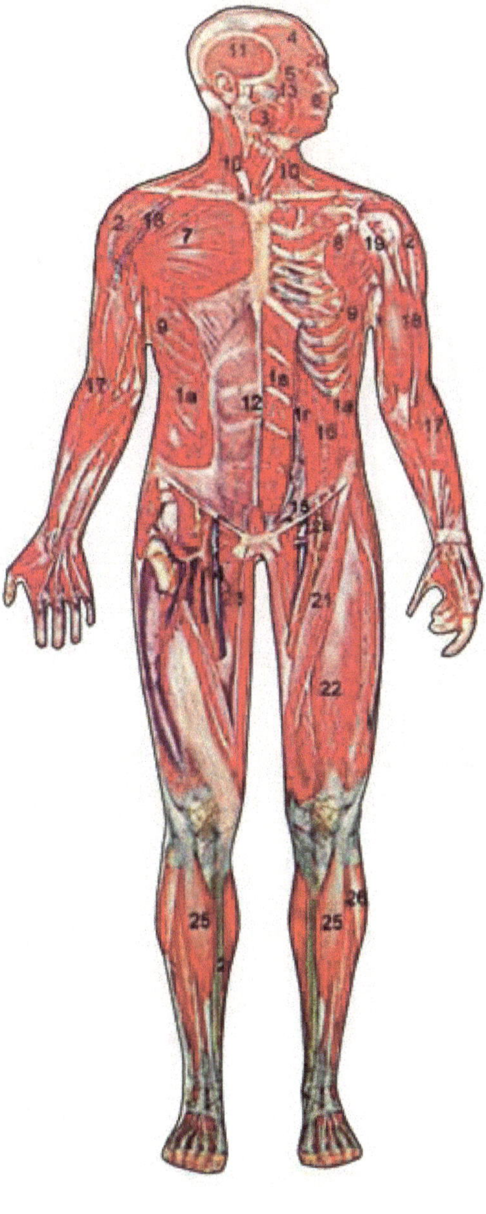

M. obliquus externus (1a)
M. obliquus internus (16)
M. rectus abdominis (1e)
M. transversus abdominis (1c)
M. deltoideus (2)
M. masseter (3)
M. epicranius (4)
M. orbicularis oculi (5)
M. orbicularis oris (6)
M. pectoralis major (7)
M. pectoralis minor (8)
M. serratus anterior (9)
M. sternocleido-mastoideus (10)
M. temporalis (11)
umbillicus (12)
M. zygomaticus major (13)
A. et V. femoralis (14)
Canalis inguinalis (15)
M. biceps brachii (16)
M. brachioradialis (17)
V. cephalica (18)
M. coracobrachialis (19)
Os. nasale (20)
M, sartorius (21)
M, quadriceps femoris (22)
M, adductor magnus (23)
M. tibialis anterior (24)
M. peroneus longus (25)
M. soleus (26)
M. gastrocnemius (27)
N. femoralis (28)

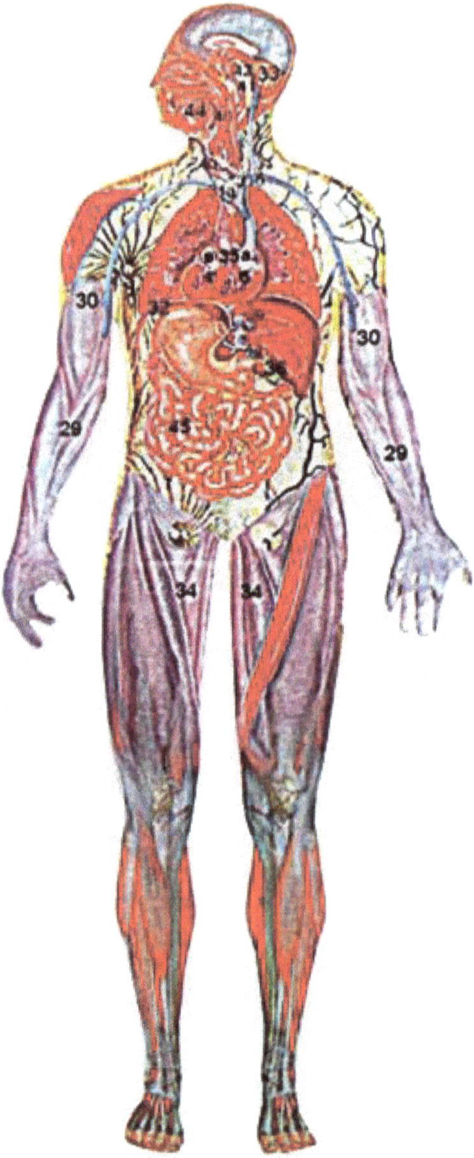

Fascia anterbrachii (29)
Fascia brachii (30)
V. brachiocephalica sinistra (31a)
V. brachiocephalica dextra (31b)
Diaphragma (32)
Cerebellum (33)
Fascia lata (34)
Car (35)
Atrium dextrum (35a)
Ventriculus dexter (35b)
Atrium sinistrum (35)
Ventriculus sinistrum (35)
Vesica fellea (36)
V. hepatica (39)
V. jugularis interna (40)
Medulla oblangata (41)
Pons (42)
V. portae (43)
Lingue (44)
Intestinum tenue (45)
Pharinx (46)

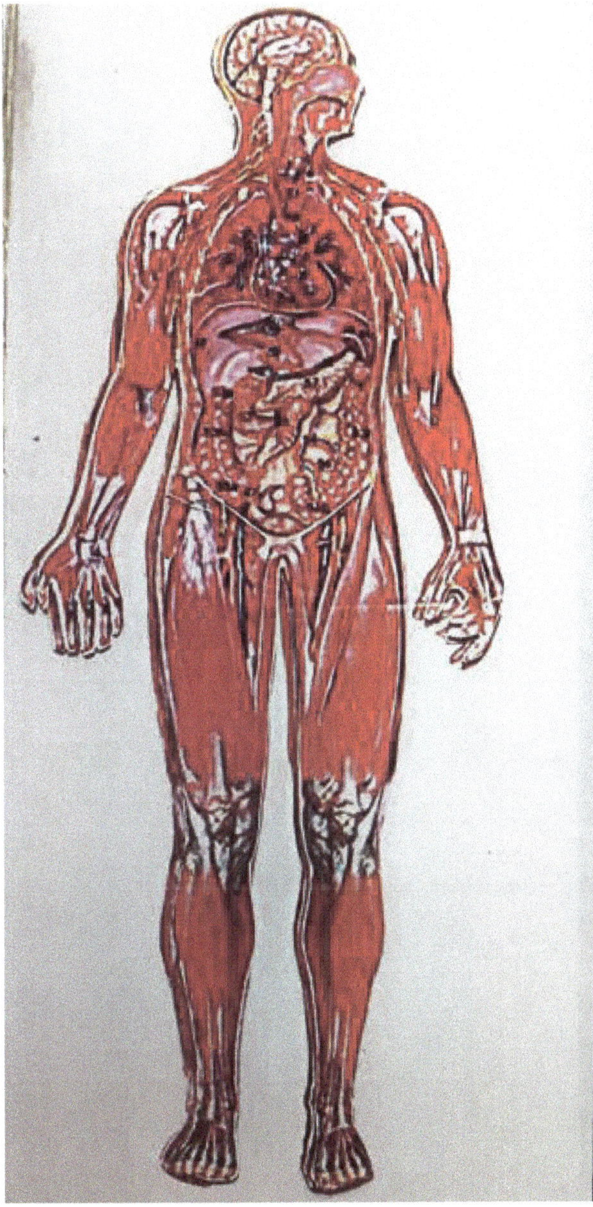

Appendix vermiformis (47)
Aorta (48)
A. brachialis (49)
Truncus coeliacus (50)
A. carotis communis (51)
Duadenum (52)
Interstinum crossum
Caecum (53a)
Colon ascendens (53b)
Colon transversum (53)
Colon dedcendens (53)
Colon sigmoideum (53)
A. mesenterica inferior (54)
A. mesenterica superior (55)
Mesenterium (56)
Pancreas (57)
Peritoneum parictale (58)
A. pulmonalis (59)
V. pulmonalis (60)
Splen (61)
A.et V. lienalis (62)
V. cava superior (63)
V. cava inferior (64)

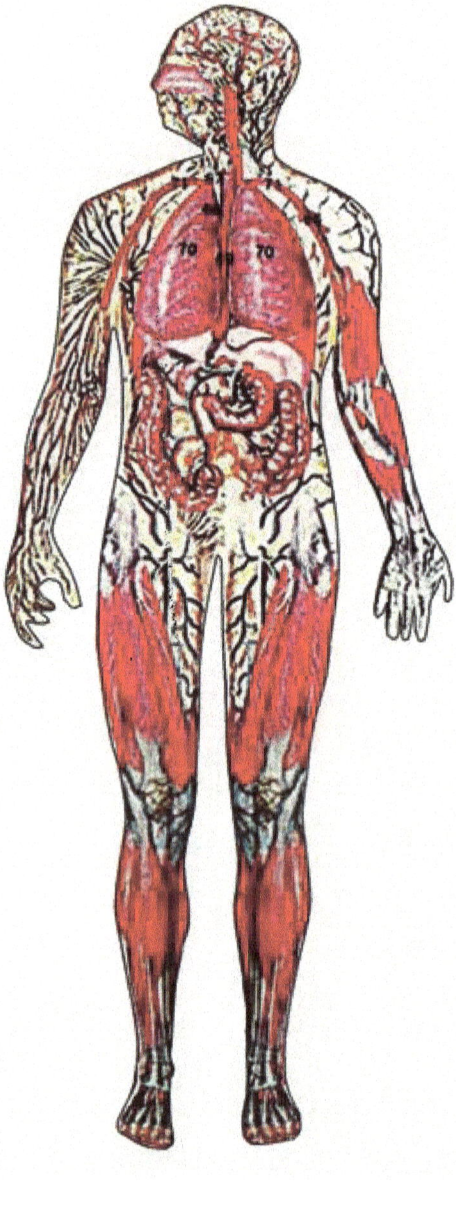

A. axillaris (65)
Truncus brachiocephalicus (66)
Trachea (67)
Ductus choledochus (68)
Esophagus (69)
Pulmones (70)
A. subciavia (71)

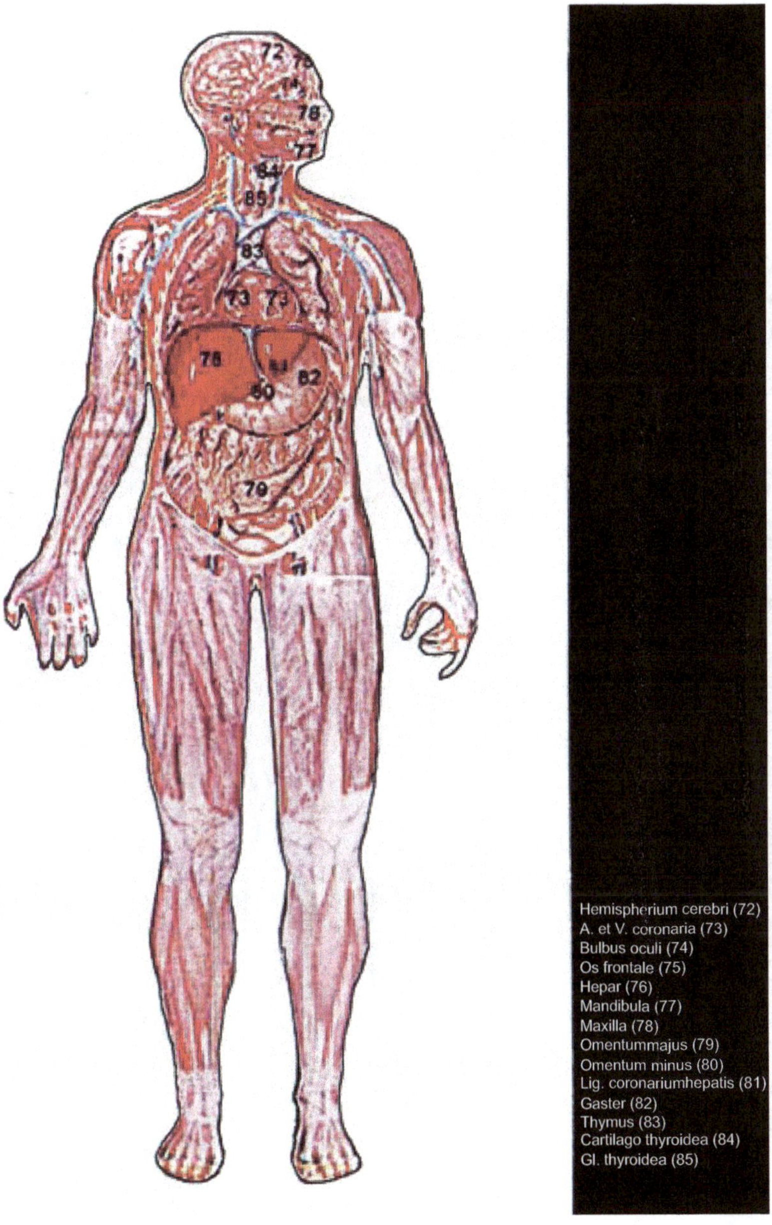
Hemispherium cerebri (72)
A. et V. coronaria (73)
Bulbus oculi (74)
Os frontale (75)
Hepar (76)
Mandibula (77)
Maxilla (78)
Omentummajus (79)
Omentum minus (80)
Lig. coronariumhepatis (81)
Gaster (82)
Thymus (83)
Cartilago thyroidea (84)
Gl. thyroidea (85)

Clavicula (86)
A. et V. epigastrica
Inferior (87)
Orbita (88)
Mm.intercostales (89)
Nn.et Vv.intercostales (90)
Costa (91)
Cartilage costae (92)
Gl. parotidea (93)
Gl. sublingualis (94)
Gl. submandibularis (95)
Sinus frontalis (96)
Sinus maxillaris (97)
Sternum (98)
M. transversus thoracis (99)

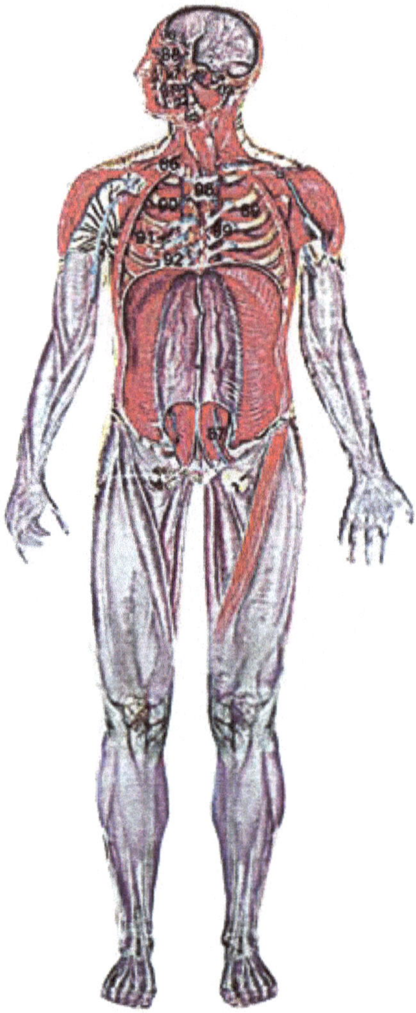

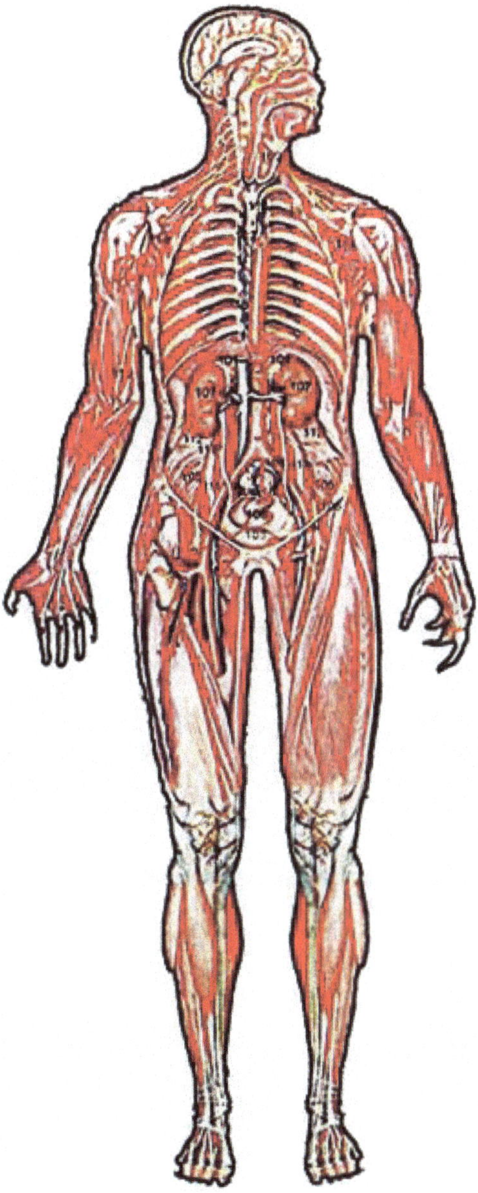

Glandula suprarenalis (101)
V. azygos (102)
Vesica urinaria (103)
A. et V. iliaca communis (104)
M. iliacus (105)
Lig. inguinale (106)
Ren (107)
Rectum (108)
Larynx (109)
N. medianus (110)
M.psoas major (111)
M. quadratus lumbarum (112)
A. et V. renalis (113)
M. subscapularis (114)
Truncus sympathicus (115)
Ductus thoracicus (116)
Ureter (117)

Appendix B

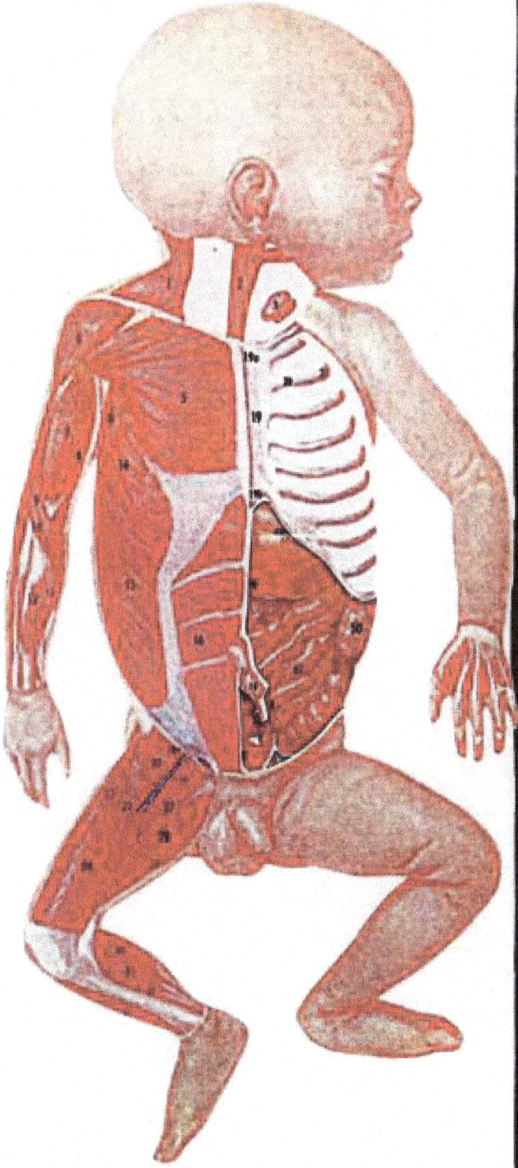

M. trapezius (1)
M. sternocleido-mastoideus (2)
Glandula thyreoidea (3)
M. deltoideus (4)
M. pectoralis major (5)
M. latissimus dorsu (6)
M. biceps brachii (7)
M. triceps brachii (8)
M. brachioradialis (9)
M. externsor carpi radialis longus (10)
M. pronator teres (11)
M. flexor digitorum superficialis (12)
M. extensor radialis brevis (13)
M. serratus anterior (14)
M. obliquus externus (15)
M.rectus abdominis (16)
Umbilicus (17)
A. umbilicalis (18)
A. et V. femoralis (18a)
Corpus sterni (19)
Manubrium sterni (19a)
Pricessus xiphaideus (19b)
Cartilago costalis (20)
Costae (21)
M. sartorius (22)
M. rectus femoris (23)
M. vastus medialis (24)
M. iliopsoas (25)
M. pectineus (26)
M. adductor longus (27)
M. gracilis (28)
M. semimembranosus (29)
M. gastrocnemius (30)
M. soleus (31)
M. flexor digitorum longus (32)
M. tibialis anterior (33)

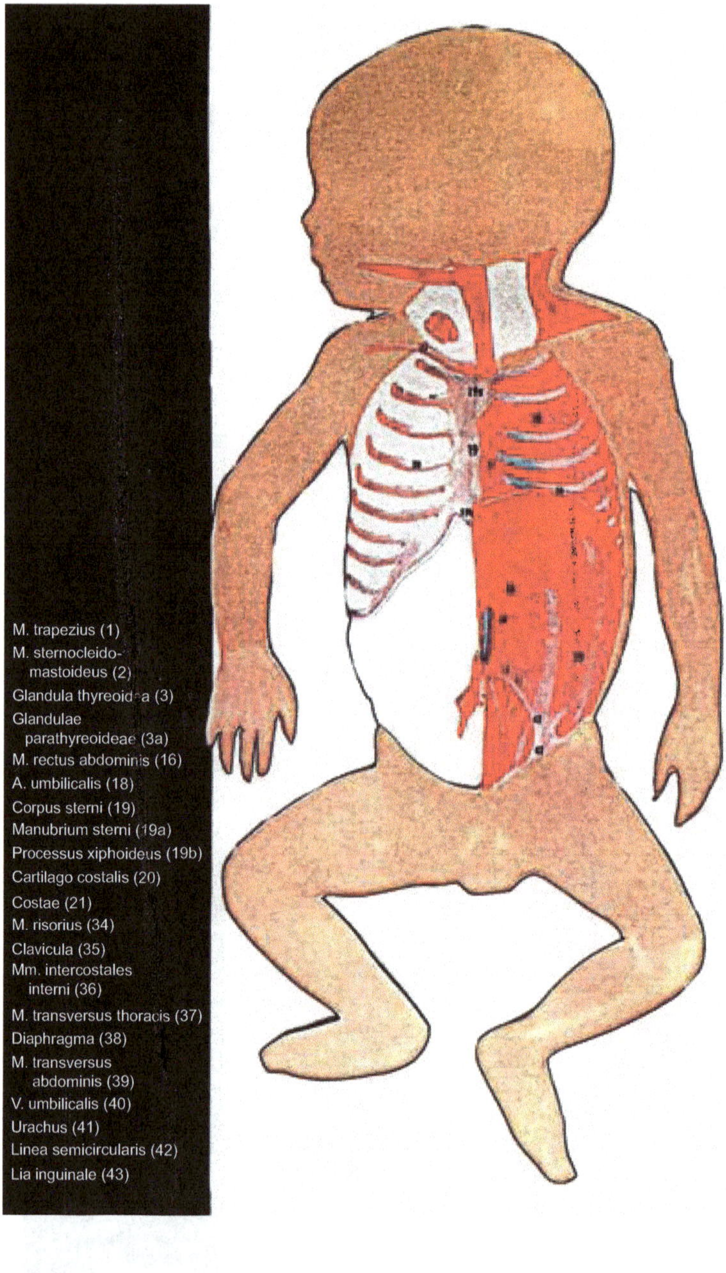
M. trapezius (1)
M. sternocleido-mastoideus (2)
Glandula thyreoidea (3)
Glandulae parathyreoideae (3a)
M. rectus abdominis (16)
A. umbilicalis (18)
Corpus sterni (19)
Manubrium sterni (19a)
Processus xiphoideus (19b)
Cartilago costalis (20)
Costae (21)
M. risorius (34)
Clavicula (35)
Mm. intercostales interni (36)
M. transversus thoracis (37)
Diaphragma (38)
M. transversus abdominis (39)
V. umbilicalis (40)
Urachus (41)
Linea semicircularis (42)
Lia inguinale (43)

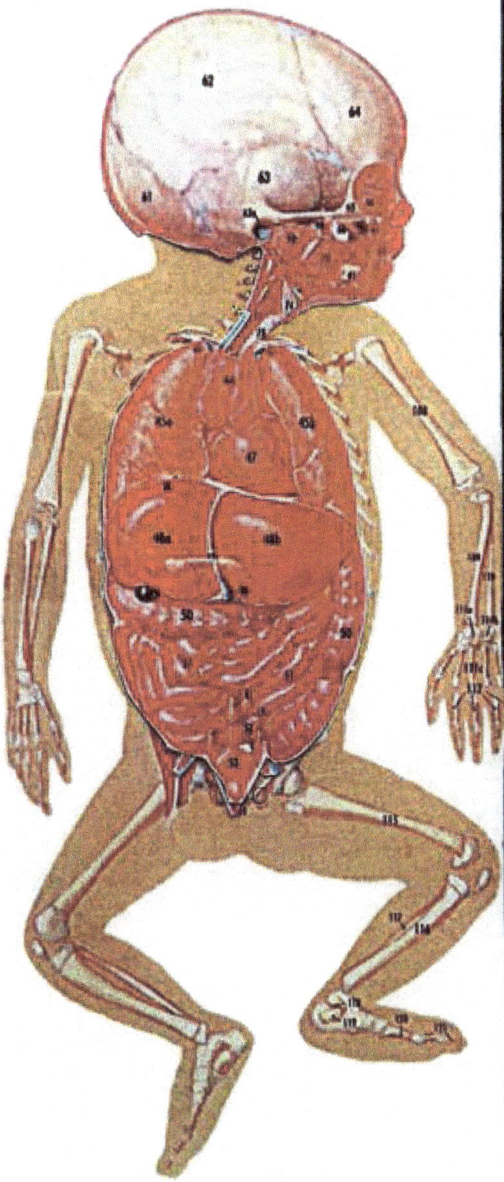

A. Umbilicalis (18)
Diaphragma (38)
V. umbilicalis (40)
Urachus (41)
Glandula thymi (44)
Pulma dexter (45a)
Pulma sinister (45b)
V. subclavia (46)
Cor (47)
Lobus hepatis dexter (48a)
Lobus hepatis sinister (48b)
Lig. falciforme hepatis (48c)
Vesica fellea (49)
Colon (50)
Jejunum et ileum (51)
Ureter (52)
Vesica urinaria (53)

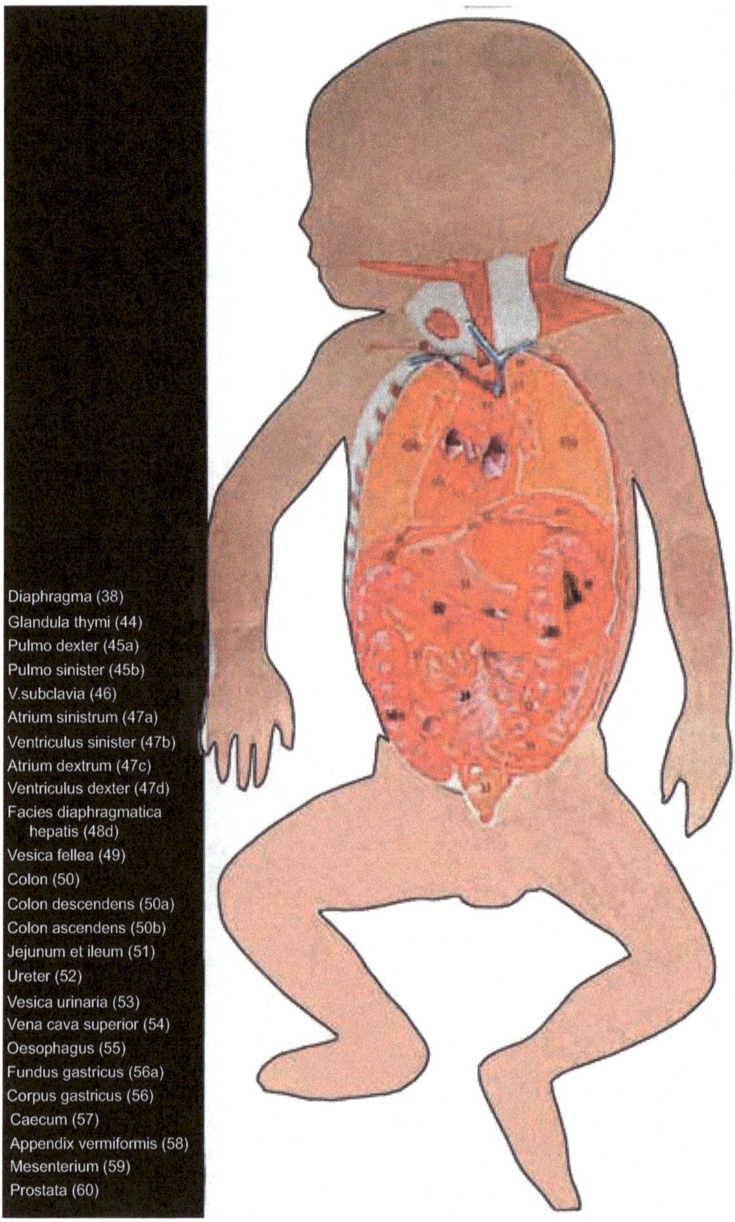
Diaphragma (38)
Glandula thymi (44)
Pulmo dexter (45a)
Pulmo sinister (45b)
V.subclavia (46)
Atrium sinistrum (47a)
Ventriculus sinister (47b)
Atrium dextrum (47c)
Ventriculus dexter (47d)
Facies diaphragmatica hepatis (48d)
Vesica fellea (49)
Colon (50)
Colon descendens (50a)
Colon ascendens (50b)
Jejunum et ileum (51)
Ureter (52)
Vesica urinaria (53)
Vena cava superior (54)
Oesophagus (55)
Fundus gastricus (56a)
Corpus gastricus (56)
Caecum (57)
Appendix vermiformis (58)
Mesenterium (59)
Prostata (60)

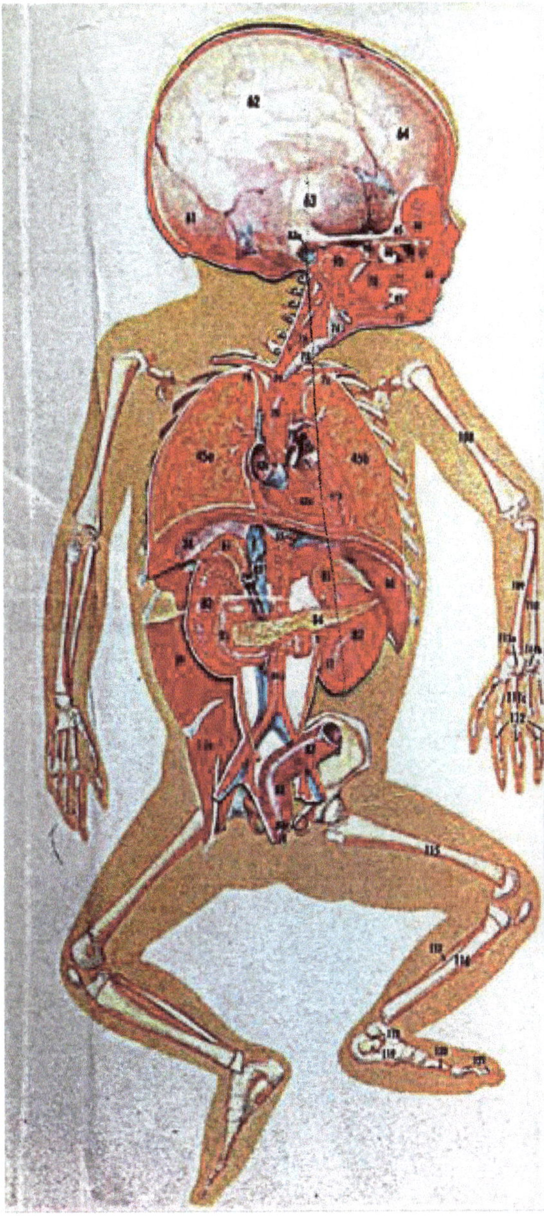

Diaphragma (38)
Pulmo dexter (45a)
Pulmo sinister (45b)
Alrium sinistrum (47a)
Ventriculus sinister (47b)
Atrium dextrum (47c)
Ventriculus dexter (47d)
Ureter (52)
Oesophagus (55)
Os occipitale (61)
Os parietale (62)
Os temporale (63)
Cavum tympani (63a)
Os frontale (64)
Os zygomaticum dextrum (65)
M. orbicularis oculi (66)
M. zygomaticus minor (67)
M. zygomaticus major (68)
M. orbicularis oris (69)
M. masseter (70)
M. buccinator (71)
M. digastricus venter anterior (72)
A. carotis communis dextra (73)
Larynx (74)
Trachea (75)
A. subclavia (76)
A. anonyma (77)
Aorta (78)
A. pulmonalis (79)
Lien (80)
Glandula suprarenalis (81)
Ren (82)
Duodenum (83)
Pancreas (84)
Vena cava inferior (85)
Aorta descendens (86a)
Colon sigmoideum (87)
Rectum (88)
Canolis analis (88a)
Porta hepatis (89)
Glandula parolis (90)
Ductus parolideus (90a)
Glandula submandidularis (91)

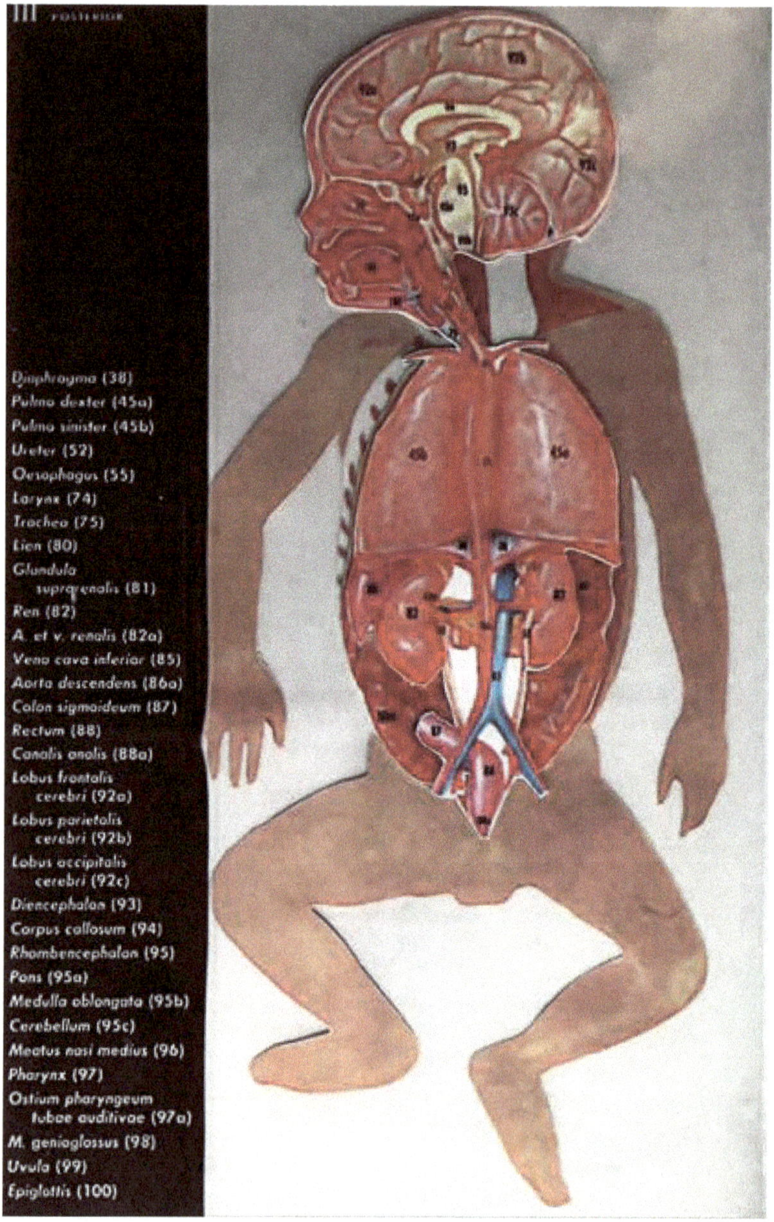
III
Diaphragma (38)
Pulmo dexter (45a)
Pulmo sinister (45b)
Ureter (52)
Oesophagus (55)
Larynx (74)
Trachea (75)
Lien (80)
Glandula suprarenalis (81)
Ren (82)
A. et v. renalis (82a)
Vena cava inferior (85)
Aorta descendens (86a)
Colon sigmoideum (87)
Rectum (88)
Canalis analis (88a)
Lobus frontalis cerebri (92a)
Lobus parietalis cerebri (92b)
Lobus occipitalis cerebri (92c)
Diencephalon (93)
Corpus callosum (94)
Rhombencephalon (95)
Pons (95a)
Medulla oblongata (95b)
Cerebellum (95c)
Meatus nasi medius (96)
Pharynx (97)
Ostium pharyngeum tubae auditivae (97a)
M. genioglossus (98)
Uvula (99)
Epiglottis (100)

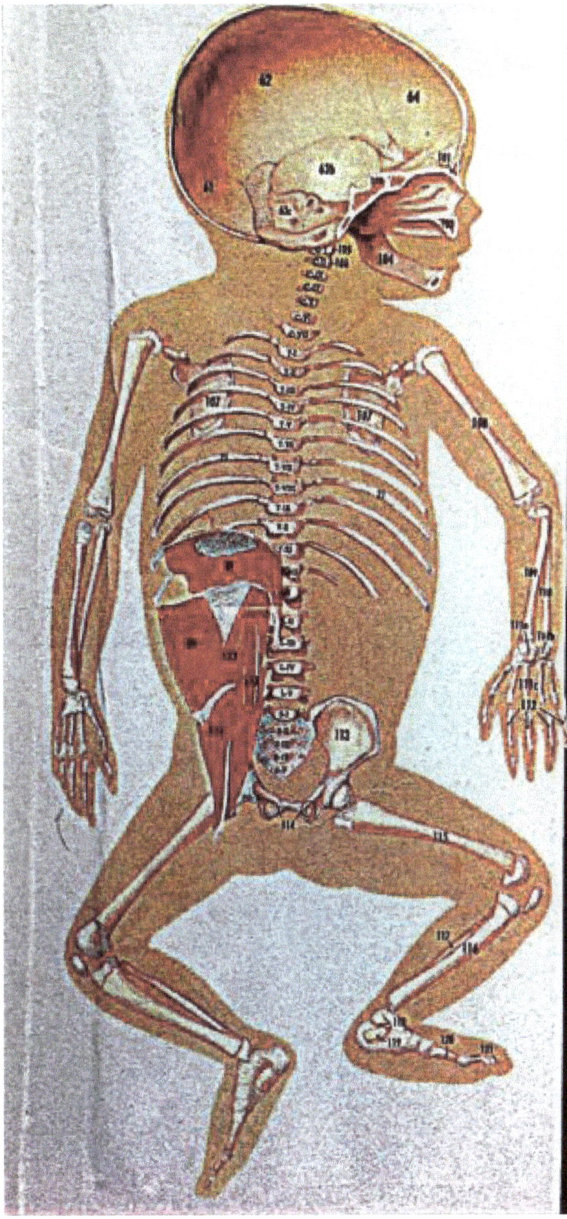

Castae (21)
Diaphragma (38)
M. transversus
abdominis (39)
Os occipitale (61)
Os parietale (62)
Pars squamosa
os temporale (63b)
Pars petrosa
os temporale (63c)
Os frontale (64)
Os nasale (101)
Os sphenoidale (102)
Os palatinum (103)
Corpus mandibulae (10
Atlas (105)
Axis (106)
Scapula (107)
Humerus (108)
Radius (109)
Ulna (110)
Os capitatum (111a)
Os hamatum (111b)
Ossa metacarpalia (111
Phalanges (112)
Os illium (113)
Os isschii
et os pubis (114)
Femur (115)
Tibia (116)
Fibula (117)
Talus (118)
Calcaneus (119)
Ossa metatarsalia (120)
Phalanges (121)
M. quadratus
lumborum (122)
M. psoas major
et minor (123)
M. iliacus (124)

Also of Interest

By the same author:

Ultrasonic Topographical and Pathotopographical Anatomy: A Color Atlas, by Z. M. Seagal and O. V. Surnina, ISBN 9781119223573. Using ultrasonic technology to create full-color detailed pictures, this atlas details the topographical and pathotopographical anatomy of the human body, as a useful reference for medical professionals and students alike. *NOW AVAILABLE!*

Check out these other titles from Scrivener Publishing:

Compendium of Biophysics, by Andrey B. Rubin, ISBN 9781119160250. The most thorough coverage of biophysics available, in a handy, easy-to-read volume, perfect as a reference for experienced engineers or as a textbook for the novice. *NOW AVAILABLE!*

Fundamentals of Biophysics, by Andrey B. Rubin, ISBN 9781118842454. The most up-to-date and thorough textbook on the fundamentals of biophysics, for the student, professor, or engineer. *NOW AVAILABLE!*

Ethics in the University, by James G. Speight, ISBN 9781118872130. Examining the potential for unethical behavior by all academic staff, both professionals and non-professionals, this groundbreaking new study uses documented examples to show where the matter could have been halted before it became an ethics issue and how to navigate the maze of today's sometimes confusing ethical academic arena. *NOW AVAILABLE!*

Reactive Oxygen Species: Signaling Between Hierarchical Levels in Plants, by Franz-Josef Schmitt and Suleyman I. Allakhverdiev, ISBN 9781119184881. Not available anywhere else, this groundbreaking volume presents a novel and unique approach to understand the complex chemical network behind the internal network structure of plants, a valuable tool for students of physics, biological physics, chemistry and biology, but also for sociologists and economists as well as for scientists in any field. *NOW AVAILABLE!*

Lightning Source UK Ltd.
Milton Keynes UK
UKHW02n1302050118
315602UK00002B/2/P

9 781119 459736